T0203103

Terence N. Mitchell · Burkhard Costisella

NMR – From Spectra to Structures

Terence N. Mitchell · Burkhard Costisella

NMR – From Spectra to Structures

An Experimental Approach

Second Revised and Expanded Edition
with 168 Figures

 Springer

Terence N. Mitchell
Universität Dortmund
– Fachbereich Chemie –
44227 Dortmund
Germany
e-mail: terence.mitchell@uni-dortmund.de

Burkhard Costisella
Universität Dortmund
– Fachbereich Chemie –
44227 Dortmund
Germany
e-mail: burkhard.costisella@uni-dortmund.de

Library of Congress Control Number: 2007924904

ISBN 978-3-540-72195-6   Springer Berlin Heidelberg New York

ISBN 978-3-540-40695-2   1st ed. Springer Berlin Heidelberg New York 2004

Springer-Verlag is a part of Springer Science+Business Media
springer.com

Cover design: WMXDesign GmhH, Heidelberg, Germany
Typesetting and production: LE-TEX Jelonek, Schmidt & Vöckler GbR, Leipzig, Germany
Printed on acid-free paper   SPIN 12028634   52/3180 YL 5 4 3 2 1 0

dedicated to Reiner Radeglia
an NMR pioneer in a then divided Germany

Preface to the Second Edition

Our attempt to present NMR spectroscopy to the beginner in a somewhat different way was well-received, so that we were invited by Springer to make some additions to the original for a second edition. Naturally we have modified the text to take account of justified criticisms of the first edition. We decided immediately to extend the number and scope of the problems section comprising Part 2, as we know that this section has been very useful to our readers. We felt that solid-state NMR is now so important and so relatively easy to do that it would be well worth giving the reader a brief account of its advantages and disadvantages. And, having already dealt with four important nuclei in some detail, we decided to add some basic information on a number of other spin-½ nuclei which are now often studied.

We thank Prof. Janet Blümel, Texas A&M University, and the Gesellschaft Deutscher Chemiker for allowing us to reproduce solid state NMR spectra. In addition we thank Klaus Jurkschat and Bernhard Lippert and their groups for making available samples of organometallic molecules. Thanks also go to Andrea Bokelmann and Bernhard Griewel for their valuable technical help.

Preface

Why write another NMR book? Most of the many already available involve theoretical approaches of various kinds and levels of complexity. Few books deal with purely practical aspects and a handful are slanted towards problem-solving. Collections of problems of different complexity are invaluable for students, since theory of itself is not very useful in deducing the structure from the spectra.

However, there is now a huge variety of NMR experiments available which can be used in problem-solving, in addition to the standard experiments which are a "must". We start by providing an overview of the most useful techniques available, as far as possible using one single molecule to demonstrate which information they bring. The problems follow in the second part of the book.

> **Readers can obtain a list of answers to the problems by application (by e-mail) to the authors**

We thank Annette Danzmann and Christa Nettelbeck for their invaluable help in recording the spectra and our wives Karin and Monika for their patience and support during the writing of the book. We also thank Bernd Schmidt for reading the manuscript and giving us valuable tips on how it could be improved. Finally, we thank the staff at Springer for turning the manuscript into the finished product you now have in your hands.

Terence N. Mitchell
Universität Dortmund
– Fachbereich Chemie –
44221 Dortmund
Germany
e-mail:
terence.mitchell@uni-dortmund.de

Burkhard Costisella
Universität Dortmund
– Fachbereich Chemie –
44221 Dortmund
Germany
e-mail:
burkhard.costisella@uni-dortmund.de

Table of Contents

Introduction

NMR spectroscopy is arguably the most important analytical method available today. The reasons are manifold: it is applied by chemists and physicists to gases, liquids, liquid crystals and solids (including polymers). Biochemists use it routinely for determining the structures of peptides and proteins, and it is also widely used in medicine (where it is often called MRI, Magnetic Resonance Imaging). With the advent of spectrometers operating at very high magnetic fields (up to 21.1 T, i.e. 900 MHz proton resonance frequency) it has become an extremely sensitive technique, so that it is now standard practice to couple NMR with high pressure liquid chromatography (HPLC). The wide range of nuclei which are magnetically active makes NMR attractive not only to the organic chemist but also to the organometallic and inorganic chemist. The latter in particular often has the choice between working with liquid or solid samples; the combination of high resolution and magic angle spinning (HR/MAS) of solid samples provides a wealth of structural information which is complementary to that obtained by X-ray crystallography. The same suite of techniques, slightly adapted, is now available to those working in the field of combinatorial chemistry. This is only a selection of the possibilities afforded by NMR, and the list of methods and applications continues to multiply.

No single monograph can hope to deal with all the aspects of NMR. In writing this book we have concentrated on NMR as it is used by preparative chemists, who in their day-to-day work need to determine the structures of unknown organic compounds or to check whether the product obtained from a synthetic step is indeed the correct one.

Previous authors have taught the principles of solving organic structures from spectra by using a combination of methods: NMR, infrared spectroscopy (IR), ultraviolet spectroscopy (UV) and mass spectrometry (MS). However, the information available from UV and MS is limited in its predictive capability, and IR is useful mainly for determining the presence of functional groups, many of which are also visible in carbon-13 NMR spectra. Additional information such as elemental analysis values or molecular weights is also often presented.

It is however true to say that the structures of a wide variety of organic compounds can be solved using just NMR spectroscopy, which provides a huge arsenal of measurement techniques in one to three dimensions. To determine an organic structure using NMR data is however not always a simple task, depending on the complexity of the molecule. This book is intended to provide the necessary tools for solving organic structures with the help of NMR spectra. It contains a series of problems, which form Part 2 of the book and which to help the beginner also contain important non-NMR information. In Part 1 a relatively simple organic compound (1) is used as an example to present the most important 1D and 2D experiments.

1

All the magnetic nuclei present in the molecule (^{1}H, ^{13}C, ^{31}P, ^{17}O, ^{35}Cl) are included in the NMR measurements, and the necessary theory is discussed very briefly: the reader is referred to suitable texts which he or she can consult in order to learn more about the theoretical aspects.

The molecule which we have chosen will accompany the reader through the different NMR experiments; the "ever-present" structure will make it easier to understand and interpret the spectra.

Our standard molecule is however not ideally suited for certain experiments (e.g. magnetic non-equivalence, NOE, HPLC-NMR coupling). In such cases other simple compounds of the same type, compounds 2–7, will be used:

2

3

4

5

6

7

Part 1: NMR Experiments

This book is not intended to teach you NMR theory, but to give you a practical guide to the standard NMR experiments you will often need when you are doing structure determination or substance characterization work, and (in Part 2) to provide you with a set of graded problems to solve. At the beginning of Part 2 we shall recommend some books which you will find useful when you are working on the problems.

We shall not attempt to present all of the many NMR experiments which have been devised by NMR experts, as this would simply make you dizzy! If at some stage you feel you want to try out other methods without ploughing through huge amounts of theory, you will find a book in the list in the Appendix which will help you to do so.

Thus we shall try to take you through Part 1 without recourse to much theory. We shall however use many terms which will be unfamiliar to you if you have not yet had a course in NMR theory, and these will be emphasized by using **bold** lettering when they appear. You can then, if you wish, go to the index of whatever theory textbook you have available in order to find out exactly where you can read up on this topic. From time to time, when we feel it advisable to say one or two words about more theoretical aspects in our text, we shall do so using *italics*.

The Appendix at the end of the book contains a list of recommended texts for theoretical and experimental aspects of NMR as well as for solving spectroscopic problems.

1
1D Experiments

1.1
^{1}H, D (^{2}H): Natural Abundance, Sensitivity

Hydrogen has two NMR-active nuclei: ^{1}H, always known as "the proton" (thus "proton NMR"), making up 99.98%, and ^{2}H, normally referred to as D for deuterium.

These absorb at completely different frequencies, and since deuterium and proton chemical shifts are identical (also because deuterium is a **spin-1 nucleus**), deuterium NMR spectra are hardly ever measured.

However, NMR spectrometers use deuterium signals from deuterium-labelled molecules to keep them stable; such substances are known as **lock substances** and are generally used in the form of solvents, the most common being deuterochloroform $CDCl_3$.

1.1.1
Proton NMR Spectrum of the Model Compound 1

Before we start with the actual experiment it is very important to go through the procedures for preparing the sample. The proton spectra are normally measured in 5-mm sample tubes, and the concentration of the solution should not be too high to avoid line broadening due to viscosity effects. For our model compound we dissolve 10 mg in 0.6 mL $CDCl_3$: between 0.6 and 0.7 mL solvent leads to optimum **homogeneity**. It is vital that the solution is free from undissolved sample or from other insoluble material (e.g. from column chromatography), since these cause a worsening of the homogeneity of the magnetic field. Undesired solids can be removed simply by filtration using a Pasteur pipette, the tip of which carries a small wad of paper tissue.

The sample is introduced into the spectrometer, locked onto the deuterated solvent (here $CDCl_3$) and the homogeneity optimized by **shimming** as described by the instrument manufacturer (this can often be done automatically, particularly when a sample changer is used).

The proton experiment is a so-called **single channel experiment**: the same channel is used for sample irradiation and observation of the signal, and the irradiation frequency is set (automatically) to the resonance frequency of the protons at the magnetic field strength used by the spectrometer.

Although some laboratories have (very expensive) spectrometers working at very high fields and frequencies, routine structure determination work is generally carried out using instruments whose magnetic fields are between 4.6975 Tesla (proton frequency 200 MHz) and 14.0296 Tesla (600 MHz). *The NMR spectroscopist always characterizes a spectrometer according to its proton measuring frequency!*

The precise measurement frequency varies slightly with solvent, temperature, concentration, sample volume and solute or solvent polarity, so that exact adjustment must be carried out before each measurement. This process, known as **tuning and matching**, involves variation of the capacity of the circuit. Modern spectrometers carry out such processes under computer control.

The measurement procedure is known as the **pulse sequence**, and always starts with a delay prior to switching on the irradiation pulse. The irradiation pulse only lasts a few microseconds, and its length determines its power. The NMR-active nuclei (here protons) absorb energy from the pulse, generating a signal.

To be a little technical: the magnetization of the sample is moved away from the z-axis, and it is important to know the length of the so-called 90° pulse which, as the name suggests, moves it by 90°, as such pulses are needed in other experiments. In the experiment we are discussing now, a shorter pulse (corresponding to a pulse angle of 30–40°, the so-called **Ernst angle**) is much better than a 90° pulse.

When the pulse is switched off, the excited nuclei return slowly to their original undisturbed state, giving up the energy they had acquired by excitation. This process is known as **relaxation**. The detector is switched on in order to record the decreasing signal in the form of the **FID** (free induction decay). You can observe the FID on the spectrometer's computer monitor, but although it actually contains all the information about the NMR spectrum we wish to obtain, it appears completely unintelligible as it contains this information as a function of time, whereas we need it as a function of frequency.

This sequence, delay-excitation-signal recording, is repeated several times, and the FIDs are stored in the computer. The sum of all the FIDs is then subjected to a mathematical operation, the **Fourier transformation**, and the result is the conventional NMR spectrum, the axes of which are frequency (in fact chemical shift) and intensity. Chemical shift and intensity, together with coupling information, are the three sets of data we need to interpret the spectrum.

Figure 1 shows the proton spectrum of our model compound, recorded at a frequency of 200 MHz (though high fields are invaluable for solving the structures of complex biomolecules, we have found that instruments operating at 200–300 MHz are often in fact better when we are dealing with small molecules).

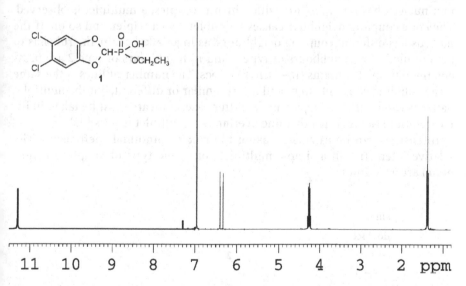

Fig. 1 Proton spectrum of compound 1 at 200 MHz. Signal assignment (from *left to right*): OH proton (singlet), aromatic protons (singlet), methine proton (doublet), OCH_2 protons (*apparently* a quintet), CH_3 protons, triplet. The small signal at 7.24 ppm is due to $CHCl_3$

Table 1 Result of a prediction compared with the actual values

Chemical shift (ppm)	J_{HP} (Hz)	Chemical shift (calc.)	J_{HP} (calc.)	Assignment
11.58	0	10.6	0	OH
6.92	not observed	7.0	0.3	CH_{arom}
6.32	28.7	6.6	16.9	CH-P
4.20	8.0	4.2	8.4	CH_2
1.33	0.6	1.3	1.0	CH_3

All signals are assigned to the corresponding protons in the molecular formula: this is made easier by prediction programmes. Table 1 presents the result of a prediction compared with the actual values.

If you do not have a prediction programme available, look on the Internet to see whether you can find freeware or shareware there. Otherwise use tables such as those you will find in the book by Pretsch et al. (see Appendix).

We shall now consider these signals and demonstrate the correctness of the assignment using different NMR techniques. First, however, some basic and important information will be provided.

The rules for **spin-spin coupling**, i.e. for determining the number of lines in a multiplet and their intensities are simple, but absolutely vital for the interpretation of any spectrum which does not just consist of a series of single lines. As far as the number of lines is concerned, the **"n+1 rule"** is applied: if a certain nucleus has n neighbours with which it couples, a multiplet is observed. Thus one coupling neighbour causes a doublet, two a triplet, and so on. If the nucleus has different coupling neighbours, as in an alkyl chain, the rule has to be modified. If n_1 neighbours of type 1 and n_2 neighbours of type 2 are present, the multiplet contains $(n_1+1)(n_2+1)$ lines. The number of lines is the same if the coupling constants to n_1 and n_2 are similar or different, but the multiplet patterns can be more complex in the latter case, and care must be taken in interpretation. Never forget that line overlap in a multiplet is possible!

Intensities can be calculated using the rule of **binomial coefficients**. The relative intensities in a simple multiplet (only one type of coupling neighbour) are as follows:

```
singlet                    1
doublet                  1   1
triplet                1   2   1
quartet              1   3   3   1
quintet            1   4   6   4   1
sextet           1   5  10  10   5   1
```

And so on. Note that in a sextet the intensities of the outer lines are very small, so that they may easily be overlooked! The same rule applies when the multiplet results from coupling to neighbours with different coupling constants (e.g. in an olefin), but more care is needed in its interpretation.

Having presented these "golden rules", we must mention that they do not always apply in this pure form. The distinction to be made here is between what spectroscopists call "first order" and "higher order" spectra. A first-order spectrum is observed when the ratio of the distance between the lines of a multiplet to the coupling constant is greater than around eight (there is no fixed boundary between first-order and higher-order spectra). Given the high fields at which modern spectrometers operate, first-order spectra are observed in the majority of cases.

When the ratio is less than around eight, changes occur in the resulting multiplet. As the ratio decreases, the intensities of the lines begin to change: the outer lines become weaker and the inner lines stronger, though the number of lines does not change. The multiplets also become asymmetric, as you will see in Fig. 1.

Even smaller ratios lead to drastic changes in the spectra, which are discussed in detail in many NMR textbooks. This should not worry you at this stage, but it is advisable to point out that spectra of aromatic groups (substituted or unsubstituted) may often not be easy to interpret because the chemical shifts are so similar.

Turning to the spectrum in Fig. 1, let us start with the one-line signal on the left, the singlet, at 11.58 ppm. Our standard, tetramethylsilane TMS, gives a one-line signal whose chemical shift is defined as 0.00 ppm. Signals to its left are said to absorb at lower field (the traditional term: many authors now use the expression "higher frequency"), those to its right (quite unusual in fact) at higher field (lower frequency) than TMS. Thus the signal at 11.58 ppm is that which absorbs at the lowest field, and we have assigned this as being due to the OH-proton. This proton is acidic, the O–H bond being relatively weak, and can thus undergo fast chemical exchange with other water molecules or with deuterated water, D_2O. Thus if our sample is treated with 1–2 drops of D_2O and shaken for a few seconds the OH signal will disappear when the spectrum is recorded again: a new signal due to HOD appears at 4.7 ppm.

This technique works for any acidic proton present in a compound under investigation and is very useful in structure determination.

The next signal is a very small one at 7.24 ppm and comes from the small amount of $CHCl_3$ present in the $CDCl_3$.

The singlet at 6.92 ppm is due to the two aromatic protons: these have identical environments and thus show no coupling with other protons. They are too far from the phosphorus atom to show measurable coupling to it.

The two lines between 6.25 and 6.40 ppm are in fact a doublet due to the methine (CH) proton, which absorbs at relatively low field because it is bonded to two electronegative oxygen atoms. This proton is very close (separated by only two bonds) to the phosphorus, which is a **spin-½ nucleus** (there is only one isotope, phosphorus-31). The proton is also a spin-½ nucleus, so that H–H and H–P coupling behaviour is analogous. The distance between the two lines in the doublet is the coupling constant J, or to be exact $^2J_{P-C-H}$ and must be given in Hz, *not* ppm! The actual J value is 28.7 Hz.

How can we show that the two lines are due to a coupling? We need to carry out a so-called **decoupling** experiment, which "eliminates" couplings. Since two different nuclei are involved here, we do a **heterodecoupling** experiment (as opposed to **homodecoupling** when only one type of nucleus is involved, most commonly the proton). Decoupling is a 2-channel experiment in which we excite (and observe) the protons with channel 1 and excite the phosphorus nuclei with channel 2, which we call the decoupling channel. Channel 2 is set to the phosphorus resonance frequency, which we can obtain from tables; the excitation of the phosphorus eliminates the coupling. Figure 2 shows the sig-

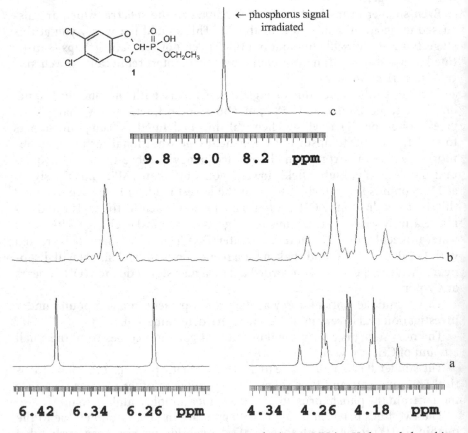

← phosphorus signal irradiated

Fig. 2a–c Heterodecoupling experiment on compound 1 (at 200 MHz). **a** Undecoupled methine and methylene signals; **b** signals after decoupling of the phosphorus. **c** ^{31}P spectrum, showing the signal which is irradiated using the decoupling channel (channel 2)

nals due to the CH proton (ca. 6.3 ppm) and the OCH$_2$ protons (ca. 4.2 ppm) before (lower traces) and after (upper traces) decoupling. The top trace shows the ^{31}P signal which is irradiated. On irradiation, the methine doublet is transformed to a singlet, the chemical shift of which lies exactly at the centre of the initial doublet.

The OCH$_2$ signal at ca. 4.2 ppm in the undecoupled spectrum consists of 8 lines and is due to those methylene protons which have only one oxygen atom in their neighbourhood rather than two. Heterodecoupling reduces the number of lines to 4; we now have a quartet with line intensities 1:3:3:1; thus phosphorus couples with these methylene protons across 3 bonds ($^3J_{P-O-C-H}$). The quartet in the decoupled spectrum (upper trace) is due to coupling of the CH$_2$ protons with the three equivalent CH$_3$ protons ($^3J_{H-C-C-H}$): this can be demonstrated by a homodecoupling experiment, a further 2-channel experiment where the second channel is used for *selective* irradiation of the methyl proton signal (a triplet, intensity 1:2:1) at 1.33 ppm (the only signal we have not yet discussed). The result is now the elimination of ($^3J_{H-C-C-H}$). leading to a doublet signal, the distance between the lines being equal to ($^3J_{P-O-C-H}$).

Thus the original 8-line multiplet is a doublet of quartets (dq).

We can now use a homodecoupling experiment to show that in the methyl signal (triplet, with each line split into a doublet) at 1.33 ppm, the distances between lines 1 and 3, 2 and 4, 3 and 5 or 4 and 6 are equal to ($^3J_{H-C-C-H}$): we irradiate the methylene protons and observe the methyl protons. The result of this experiment is shown in Fig. 3.

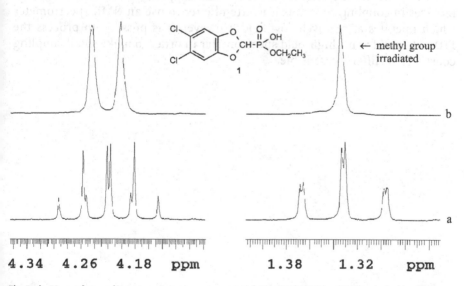

← methyl group irradiated

b

a

4.34   4.26   4.18   ppm   1.38   1.32   ppm

Fig. 3a,b Homodecoupling experiment on compound 1 (at 200 MHz). a Undecoupled methylene and methyl signals; b signals after irradiation of the methyl group

Below we see the signals due to OCH_2CH_3 on the left and OCH_2-CH_3 on the right. After decoupling (above), the 8-line OCH_2CH_3 signal becomes a doublet due to the P–H coupling, which is of course still present. The 6-line OCH_2-CH_3 signal, the one which is irradiated, becomes one single line. This experiment was carried out on a state-of-the-art spectrometer: earlier spectrometers would more likely have shown the decoupled OCH_2-CH_3 signal in a highly distorted form.

Homo- and heterodecoupling experiments such as those described here are used routinely in structural analysis and can be carried out very rapidly. In the present case they have provided exact proof that the signal assignments were correct.

1.1.2
Field Dependence of the Spectrum of 1

The decoupling experiments which we have just discussed showed that the multiplet (doublet of quartets) due to the OCH_2 group arises from the presence of two coupling constants which are of similar magnitude ($^3J_{HH}$ 7.1 and $^3J_{POCH}$ 8.0 Hz). We could see all 8 lines clearly in the spectrum, which was measured at 200 MHz. If we compare this multiplet with the corresponding signals recorded at 400 and 600 MHz (Fig. 4) we do not see the eight lines so clearly.

This is easy to understand, if we remember that 1 ppm on the chemical shift axis corresponds to 200, 400 and 600 Hz respectively for the three spectrometers. Thus at higher field the multiplet appears "compressed".

Thus in fact for the determination of small coupling constants or small differences in coupling constants it is often better to use an NMR spectrometer which operates at relatively low field. However, it is possible to process the FID obtained from a high-field spectrometer in order to make small coupling constants or differences visible.

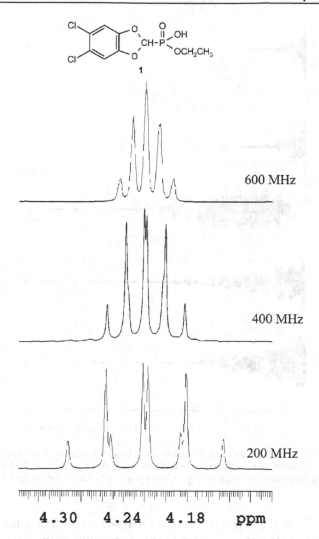

Fig. 4 OCH$_2$ proton signal of compound 1, measured using 200, 400 and 600 MHz spectrometers

1.1.3
FID Manipulation: FT, EM, SINE BELL (CH$_2$ Signal of 1)

The signal (FID, free induction decay) resulting from an NMR experiment contains the original data which are stored in the computer, and after the **Fourier transformation** (FT) we obtain the NMR spectrum itself.

We can manipulate the FID mathematically in various ways *before* Fourier transformation, in order to optimize the spectrum with respect to the **linewidth** or the **lineshape**.

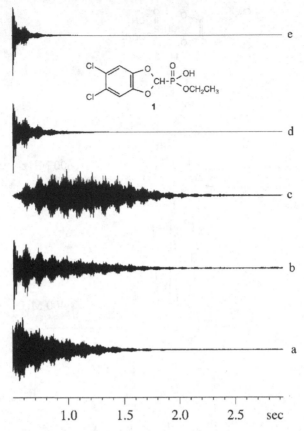

Fig. 5a–e FID of compound 1. **a** Original data; **b** multiplied by a negative line broadening function (–0.3 Hz); **c** multiplied by a shaped sine bell function (SSB = 1); **d** multiplied by a positive line broadening function (0.8 Hz); **e** multiplied by a positive line broadening function (1.9 Hz)

Figure 5 shows the original FID and the result when this is multiplied by mathematical functions: either **exponential multiplication** (EM) or **shaped sine bell** (SSB, a sine function).

EM affects the linewidth and is often also known as a **line broadening** function LB. A positive value of LB (here 0.8 and 1.9 Hz) broadens the lines, a negative value (here –0.3 Hz) sharpens them: however, never forget that we are only *modifying* the information present, so that a decrease in the linewidth is automatically accompanied by an increase in the **baseline noise**. This becomes clear immediately when we see the spectra of the OCH_2 multiplet shown in Fig. 6.

Fourier transformation without data manipulation leads to the multiplet at the bottom (a), which shows more fine structure when a negative LB value is used (b). The spectrum in the middle (c) results from use of the SSB function, and now all eight lines are clearly visible as the linewidth is much smaller. The price we pay is that the lineshape is completely changed, the positive central

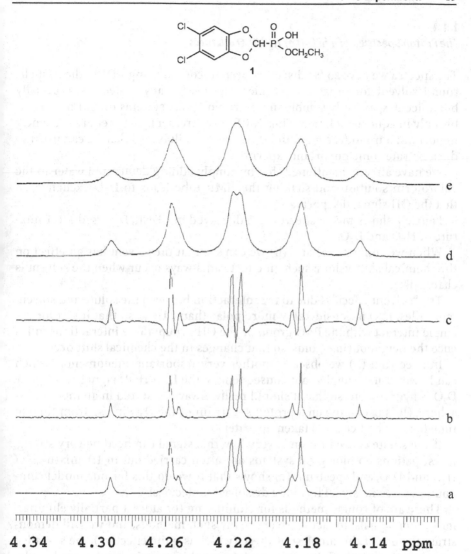

Fig. 6a–e OCH2 signal of compound 1 (200 MHz): a Only Fourier transformation; b Fourier transformation preceded by multiplication of FID by a negative line broadening function (–0.3 Hz); c Fourier transformation preceded by multiplication of FID by a shaped sine bell function (SSB = 1); d Fourier transformation preceded by multiplication of FID by a positive line broadening function (0.8 Hz); e Fourier transformation preceded by multiplication of FID by a positive line broadening function (1.9 Hz)

"real" lines being accompanied by negative "wings". Positive line broadening functions decrease the quality of the spectra considerably, but there is an improvement of the **signal to noise ratio** (d, e).

The use of sine or cosine functions in FID data processing is an essential tool in 2D NMR.

1.1.4
The Proton Spectrum of 1 in D₂O or H₂O/D₂O Mixtures

The spectra we have so far discussed were recorded using $CDCl_3$, the best all-round solvent for organic molecules. However, many molecules, especially biomolecules, are only soluble in water; biological systems often remain stable only in aqueous solution. Thus NMR measurements in water are extremely important: our model compound is also water-soluble, so that we can use it to demonstrate some important experiments.

We have already mentioned that by simply adding deuterated water to the chloroform solution and shaking the NMR tube leads to H-D exchange, so that the OH signal disappears.

Figure 7 shows the ^1H spectra of 1 dissolved in $CDCl_3$, D_2O, and a 1:1 mixture of H_2O and D_2O.

When we compare (a) and (b) we can see that the solvent has an effect on the chemical shift values; such an effect can always occur when the solvent is changed!

The "solvent effect" is due to the interaction between the solute and solvent molecules. D_2O is considerably more polar than $CDCl_3$, so that it can for example interact with the P=O group or the OH group; these interactions influence the neighbouring atoms, so that changes in the chemical shift occur.

In spectrum (b) we observe another very important phenomenon, which can however have unpleasant consequences: the H_2O/HOD signal at 4.7 ppm. D_2O is hygroscopic, so that it should really always be stored in an inert atmosphere. (It is useful to run a proton spectrum of the D_2O in use from time to time to see whether it has taken up water).

If the solute concentration is very low, this signal can become very strong; investigations on biological systems are often carried out in 1:1 mixtures of H_2O and D_2O, and spectrum (c) shows that if we do this for our model compound we see no signal from the dissolved molecules!

There are of course methods for eliminating (or at least partially eliminating) water signals; in fact there are many such methods, and we will demonstrate the use of the simplest of these (which is quite effective), the so-called **presaturation** method. Before carrying out this experiment we need to determine the exact chemical shift of the water signal which we wish to suppress using a standard proton experiment (the computer software can help us here).

Now comes the actual presaturation experiment, in which the water signal is irradiated for 1–2 sec using a pulse set to its chemical shift. This saturates the signal, which is thus no longer visible when the pulse is switched off, and only slowly regains its natural magnitude via **relaxation**. (We shall return to relaxation later).

Now we use a normal proton pulse to excite the solute molecule; spectrum (d) shows the result of the presaturation experiment carried out on the H_2O/D_2O solution of model compound 1. A residual H_2O/HOD signal can be ob-

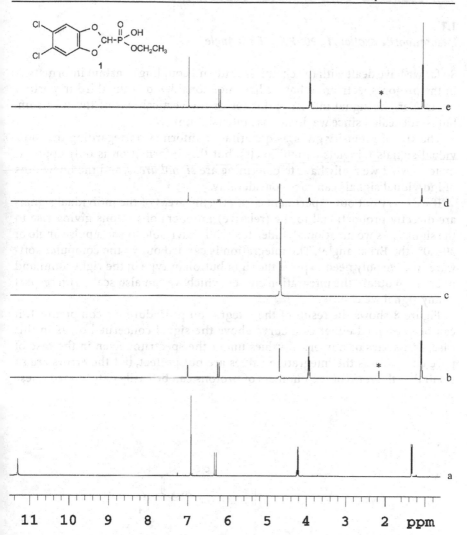

Fig. 7a–e Proton spectra of 1: a Dissolved in CDCl$_3$; b in D$_2$O; c in D$_2$O/H$_2$O; d with presaturation of the water signal; e with presaturation using a **digital filter**. Signals marked with * are due to an impurity (solvent from recrystallization of 1)

served as well as a signal due to the presaturation, but the signals of 1 can be readily seen.

We can improve the appearance of the spectrum by applying a so-called **digital filter**; the result is shown in spectrum (e).

One thing we can *not* prevent when carrying out presaturation or other water suppression experiments is the distortion or disappearance of solute signals which are very close to (within a few Hz of) the HOD signal!

1.1.5
Integration: Relaxation, T₁, 90°-Pulse, Ernst Angle

So far we have dealt with the chemical shift and coupling constant information in the proton spectrum. What we have not considered is the third important parameter, the signal intensity; this forms the vertical axis of the spectrum, but is not scaled since we do not use intensity units.

The signal intensity gives us quantitative information regarding the individual signals (singlets or multiplets), but this information is only approximate as what we really have to determine are *signal areas*, and the linewidths of individual signals can vary considerably.

If we carry out our experiment correctly, the areas of the individual signals are directly proportional to the (relative) numbers of protons giving rise to the signals. As we mentioned under 1.1.1, it is advisable to use a pulse angle of 30–40° (the Ernst angle). The integration is carried out by the computer software, and we only need to press the right button or type in the right command in order to obtain the integration curves, which we can also scale with respect to any signal we choose.

Figure 8 shows the result of the integration procedure for compound 1: it can be presented either as a curve above the signal concerned or, as in this case, as a series of numerical values under the spectrum. Even in the case of pure compounds the integration values are not perfect, but the errors are so small that the *ratio* of the numbers of protons can be easily determined; these

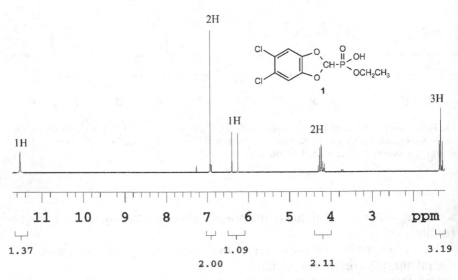

Fig. 8 Proton spectrum of 1 with (*below*) integration values and (*above*) numbers of chemically shifted protons in the molecule. The singlet due to the aromatic protons has been set equal to two protons

numbers are extremely helpful in the structure determination process. Thus here the relative numbers of protons present are given above the individual signals, while the integration values (set with respect to the aromatic protons) are given below the spectrum.

The question arises as to why the integration values are not completely accurate. One reason may be that some multiplets present are too close together, so that the software cannot find the baseline between them. However, there are also systematic errors involved, and this has to do with the **relaxation** phenomenon we mentioned above.

At the very beginning of our discussion in 1.1.1, we mentioned that any **pulse experiment** begins with a delay period. This is necessary so that the spins can return to equilibrium before they are excited. After excitation (when the pulse is turned off) we observe the FID, the free induction decay. What "decays"? The induced magnetization of the spins, and this process is known as relaxation. It may be slow or fast, as we shall see, and can also occur via a number of processes, which are discussed in detail in the monographs we have recommended for further reading. We will only treat relaxation very briefly here.

We stated previously that the signal induced by a single pulse is largest if we use a so-called **90° pulse**. When the 90° pulse is switched off, the spins "relax", and the time they need to return to equilibrium is obviously longer than if we use a shorter pulse. But a shorter pulse gives us less signal, and so the Ernst angle is a compromize. The time the spins need to return to equilibrium is called the **relaxation time**, and what we need to talk about here is the so-called **spin-lattice relaxation time** T_1 (we are dealing with liquids here, not crystals, and the term "lattice" refers to the local environment of the spins.

In order to design our experiment properly we need to have some idea of how long this T_1 is; relaxation is in fact an exponential process.

T_1 values can be easily determined using **pulse sequences** which form part of the standard computer software, the most common one being the so-called **inversion-recovery experiment**.

This experiment uses two pulses, 180° and 90°, separated by a delay time τ which is varied. For each delay a certain number of FIDs are accumulated; the result is a series of spectra in which the individual signals have different intensities. Figure 9 shows the result of an inversion-recovery experiment carried out on 1.

We can see at once that each proton behaves differently, because it has its individual relaxation time T_1; depending on the delay signals may be negative, positive, or have zero intensity. The T_1 values can be computed using spectrometer software.

One of the textbooks in our list of recommended reading states that proton T_1 values in high-resolution NMR lie close to 1 second and vary little with the type of proton.

We have carried out T_1 measurements for model compound 1 at three different frequencies (300, 400, 500 MHz). The values for the various proton sig-

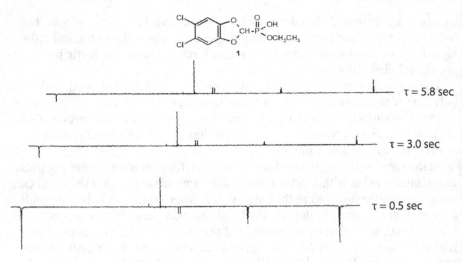

Fig. 9 Spectra of compound **1** obtained from an inversion-recovery T_1 experiment. Pulse sequence: fixed delay – 180° pulse – variable delay τ – 90° pulse – acquisition of FID

nals are shown in Table 2, while Fig. 10 shows one example of the data obtained and includes the equation used for the T_1 calculations.

Our data show that the T_1 values are generally larger than 1 second and vary drastically from signal to signal; they do not appear to vary systematically with the spectrometer magnetic field.

Since the integration values form such an important element of structure determination, we need to set the spectrometer up properly before carrying out the NMR experiment. And one very important parameter which is often forgotten is the **relaxation delay**, the delay between the single NMR experiments which allows the nuclei to relax. Remember that relaxation is an exponential process, so that theory suggests that it is necessary for the best results to set this equal to at least *five times* T_1 (in our case more than 25 sec for the aromatic protons!). The other parameter we need to set correctly is of course the pulse angle, and the following set of experiments show how these are interrelated.

We carried out two sets of experiments in which we set the pulse angle first at 90°, then at 30°. Using these two values we then varied the relaxation delay. Since the greatest difference in the relaxation times is that between the OH proton and the aromatic protons, we show in Fig. 11 the comparison between the integration values of the aromatic protons (set equal to 2.0) and of the OH proton for 90° pulses and for 30° pulses. The values approach each other with a relaxation delay of 10 sec and are virtually equal for a delay of 25 sec, but the 90° pulses give values which are completely wrong if a "conventional" delay of 1–2 sec is used! On the other hand, the error is quite low if the delay is set at 2 sec and the pulse length is 30°.

Table 2 Relaxation times T_1 for the protons in compound 1 at 26°C

Spectrometer frequency (MHz)	300	400	500
	T_1 (sec.)	T_1 (sec.)	T_1 (sec.)
OH	0.4	0.5	0.6
Aromatic H	5.4	5.3	5.6
CH	3.3	3.3	3.6
OCH$_2$	2.7	2.8	2.9
CH$_3$	2.9	2.8	3.0

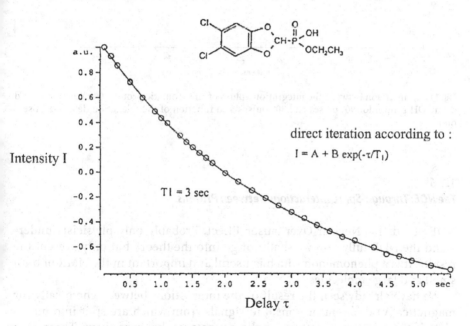

direct iteration according to :

$$I = A + B \exp(-\tau/T_1)$$

TI = 3 sec

Intensity I

Delay τ

Fig. 10 Compound 1: T_1 determination for the methyl signal (at 500 MHz in CDCl$_3$ at 26°C). Plot of signal intensity against delay τ. The computer software gives a T_1 value of 3 sec

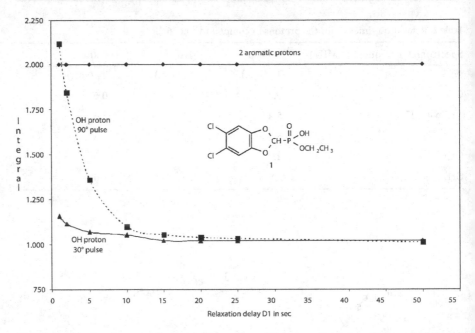

Fig. 11 Comparison between the integration values of the aromatic protons (set equal to 2) and of the OH proton for 90° pulses and 30° pulses as a function of the relaxation delay D1 in seconds

1.1.6
The NOE: Through-Space Interactions between Protons

NOE stands for Nuclear Overhauser Effect. Probably only physicists understand the NOE fully, and we shall not go into the theory but only present the results. It is a phenomenon which is useful and important in the NMR of both small and large molecules.

We have already seen the result of the interactions between chemically (or magnetically) different protons, the signals from which are split into multiplets if there is a measurable coupling constant J between them. These coupling constants are the result of the so-called **scalar coupling** in which information about spin states is transferred via the bonding electrons and can be observed across several bonds, depending on the hybridization of the intermediate carbon atoms. (There is also a so-called **through-space coupling**, but this is not often observed, so that we shall not go into it in this book).

The NOE depends on a special kind of **relaxation** known as **dipole-dipole cross-relaxation**. When one signal in an NMR spectrum is irradiated, the intensities of others may change; this is called the NOE and its importance is due to the fact that the signals which react are due to spins which are physically close to that perturbed by irradiation.

We have previously used signal irradiation to *simplify multiplets*: this is the phenomenon known as decoupling (see Section 1.1.1). The NOE interactions are also demonstrated by using signal irradiation, and just as in decoupling we set up the spectrometer so that just one particular proton signal is affected. When we irradiate this signal, we are of course feeding energy into the **spin system**, thus displacing it from equilibrium: the system tries to get back to equilibrium by using relaxation processes involving dipole-dipole cross-relaxation, and the visible result is changes in signal intensity. These can be positive or negative, depending on (among other things) the size of the molecule: for small molecules they are positive, but for molecules with a molecular weight larger than about 2 kD they are negative. The change of the signal intensity is known as the NOE.

These remarks only apply to the proton-proton NOE; experiments involving an NOE between the proton and another nucleus can also be carried out, and the NOE also has an effect on certain carbon-13 spectra, as we shall see later.

Theory tells us that the maximum gain in proton signal intensity is 50%, but normally we are dealing with changes of only a few per cent, and the magnitude of these is dependent on the distance between the irradiated proton(s) and the observed ones; the effect is too small to be visible when this distance exceeds about 5 Å.

*The NOE is really quite complicated, and in fact even small molecules can show negative NOEs, which are due to a phenomenon known as **spin diffusion**.*

Why is the NOE so important to the NMR spectroscopist? Because it allows us to obtain information about the 3-dimensional structure of the molecule under consideration *in solution* (remember: the only other way to do this is by X-ray structural analysis, but this only works for substances which give good-quality crystals, and by definition not for liquids). Thus we can obtain information on conformations or configurations, something which is particularly important for biomolecules such as proteins, where NOE measurements are absolutely vital.

There are two-dimensional NOE experiments (see below, Section 2.3), but first we shall consider the one-dimensional measurements, which are of two types. To make these clear we shall use molecules **1** and **3**.

1.1.6.1
NOE Difference Spectroscopy

Here we record two proton spectra alternately, one the normal one and the other that in which we irradiate one of the signals. The first spectrum contains no NOE information, while the second does. The resulting FIDs are subtracted from one another by the computer, and the result is a spectrum in which only those signals are present for which intensity differences are observable.

Figure 12b shows such an NOE difference spectrum for the acetal **3**; the spectrum was obtained by irradiating the methine doublet at about 5.8 ppm (the normal spectrum of **3** is shown in Fig. 12a).

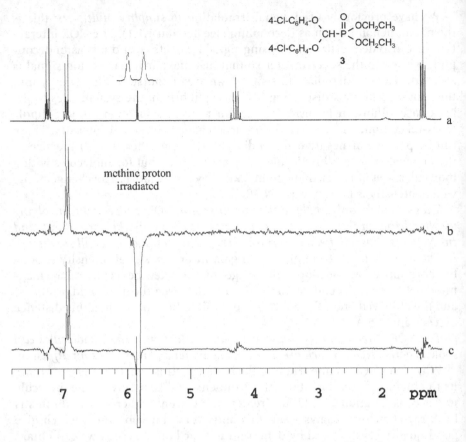

Fig. 12a–c NOE experiments carried out at 200 MHz on compound **3**. a Normal spectrum, with expansion of methine doublet; b selective NOE spectrum, total time required 18 min; c NOE difference spectrum, total time required (preparation, measurement) 42 min

A strong negative signal is always observed at the irradiation position. The baseline of the spectrum is very uneven, and it is not possible to correct the **phase** of all the signals at the same time: this is typical of NOE difference spectra, and is due to inexact subtraction of the FIDs. However, we can see a strong positive signal for one half of the AA'BB' multiplet due to the para-substituted aromatic moiety: this positive signal must be due to the protons closer to the methine proton. No further useful information is available from this experiment, which we can compare with the second technique described below.

1.1.6.2
Selective 1D NOE Experiment (1D-NOESY) and Selective 1D TOCSY Experiment

Advances in computer and spectrometer design have made possible an NOE experiment which does not rely on spectrum subtraction. This is sometimes referred to as NOESY (Nuclear Overhauser Experiment SpectroscopY). Again we will not go into details, but this technique relies on excitation of the proton(s) to be irradiated using selective pulses (**shaped pulses** of exactly predetermined width and intensity). The result of such measurements, shown for compound **3** in Fig. 10c, is that only those signals are observed which experience a positive NOE, and thus a positive signal enhancement., or – more rarely for small molecules but always for large molecules – a negative signal enhancement (negative NOE). The baseline is now very straight, so that even small signals are clearly visible.

The same proton is irradiated, and just as in the difference experiment, one aromatic pseudo-doublet shows a strong NOE; a very weak but just visible effect is shown by the OCH$_2$ protons.

You may wonder why we did not use our model compound **1** in order to demonstrate the NOE. The reason becomes quite clear when we look at the result of a selective NOE experiment carried out at 600 MHz on **1**, which is shown in Fig. 13.

The normal spectrum is shown below, the selective NOE spectrum, again with irradiation of the methine doublet, above.

Although the structural formulae of **1** and **3** are very similar, their NOE behaviour is very different: *all* the protons of **1** show an NOE! The reasons for this become clear when we refer to the known X-ray crystal structures of **1** and **3**. Although these depict a defined arrangement in the crystal, whereas NMR spectra reflect averages of possible arrangements in solution, the intramolecular distances measured from the crystal structures do in fact correlate well with the results from the NOE measurements, as is shown in Table 3 below.

In compound **1**, all interproton distances lie in a range which would be expected to give rise to an NOE, as the experiment confirmed. In **3**, although the structural formula is very similar, only the distance between the CH proton and the neighbouring "ortho" protons lies clearly in the "NOE range". The others are close to or above 5 Å, so that only very small NOEs or none at all could be expected.

We have seen that NOE experiments are very useful and can give information on relative interproton distances in the molecule. However, we should stress that NOE experiments can be difficult to interpret because of the many factors involved in their generation.

If and when you need to concern yourself with NOEs in detail, we strongly advise reading up on them in one of the books we recommend in the Appendix.

We now want to turn to another experiment which, we must make clear at the start, does not have any relationship in theory to NOE experiments. In fact the theory is so complicated that we shall not say anything about it at all, but just refer you to one of the books in the Appendix. We are including this experiment because of its unique advantages when the spectrum has overlapping multiplets. It is called TOCSY, which stands for Total Correlation SpectroscopY (it has a second, more amusing name: HOHAHA, standing for HOmonuclear HArtmann-HAhn), and is of particular use when oligosaccharides or peptides are under study.

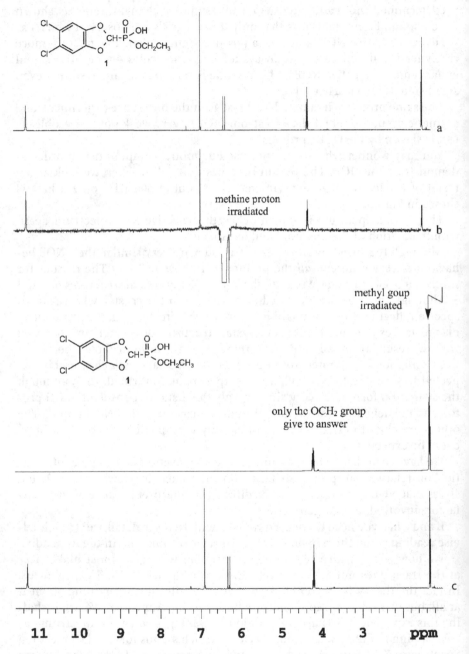

methine proton
irradiated

methyl goup
irradiated

only the OCH₂ group
give to answer

Fig. 13a–c Selective 1D NOE spectrum of 1: **a** Normal spectrum; **b** spectrum recorded with ir-radiation of the methine doublet (600 MHz, measurement time 4 min); **c** 1D TOCSY spectrum of compound 1. The methyl signal was irradiated

Table 3 Distances between the CH proton and other protons in compounds 3 and 1 (in Å)

Compound 3: distance between CH proton and		Compound 1: distance between CH proton and	
o-protons	2.07[a]	OH proton	4.04
m-protons	5.93[a]	m-protons	4.56[a]
OCH$_2$ protons	5.02[a]	OCH$_2$ protons	4.15[a]
CH$_3$ protons	4.82[a]	CH$_3$ protons	2.95[a]

[a]Shortest distance calculated

We have used compound 1 to demonstrate TOCSY, which basically tells us which multiplets in a spectrum belong to a common spin system. Thus (Fig. 13c) when the methyl signal of 1 is irradiated, there is a response ("answer") from the OCH$_2$ group *because of the coupling* between the methyl and methylene protons.

The difference between 1D NOESY and 1D TOCSY is thus the different type of interaction: in 1D NOESY through space, and by 1D TOCSY through-bond. The analogous 2D spectra are shown in Fig. 25.

1.2
^{13}C

Carbon-12, like oxygen-16, is not NMR-active. However, only 1.1% of the total carbon in a molecule consists of the spin-½ carbon-13 isotope, so that the sensitivity of this nucleus is much lower. Thus rather than using only perhaps 8 or 16 pulses, as in many proton experiments, we shall now require hundreds or even thousands of pulses, depending on the solute concentration.

1.2.1
Natural Abundance ^{13}C Spectrum of Compound 1

Organic compounds contain four types of carbon atom: methyl, methylene, methine and quaternary. And so if we simply record the spectrum as we would a proton spectrum, the result will be a series of quartets, triplets, doublets and singlets, each associated with a carbon–proton one-bond coupling constant of between 125 and 250 Hz. If we are dealing with a complex molecule, these multiplets will overlap and give us spectra which are almost impossible to analyse. In addition, coupling interactions over two or more bonds complicate the picture still further.

Thus when it became possible to record carbon-13 spectra routinely it was decided that the logical thing to do would be to decouple ALL of the protons from the carbons simultaneously (a technique known as **broad-band decoupling**) in order to obtain a carbon-13 spectrum consisting only of singlets.

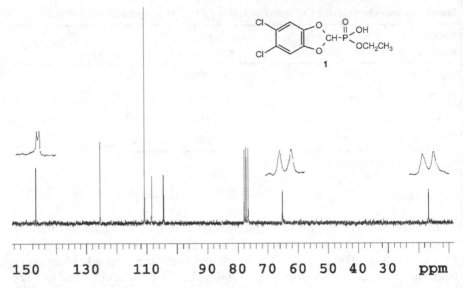

Fig. 14 Natural abundance carbon-13 spectrum of 1 (50 MHz) with expansion where necessary to show doublet structure. The assignments are as follows (*from left to right*): aromatic C bonded to oxygen (doublet) ; aromatic C bonded to chlorine (singlet); aromatic CH (singlet); methine (doublet); $CDCl_3$; OCH_2 (doublet); CH_3 (doublet). Multiplet splittings are due to coupling with phosphorus and are (except for $^1J_{PC}$) small

This gives us the chemical shift information for each type of carbon atom in the molecule. We do not have any coupling information, however, but we shall see below how we can obtain the coupling information we need.

Let us look at the natural abundance carbon-13 spectrum of our model compound 1, which is shown in Fig. 14.

If we count the number of different carbons in the molecule, we see that we expect six signals (three for the aromatic carbons, one for the methine carbon, one for the methylene and one for the methyl carbon). Each of the three aromatic carbon signals corresponds to two carbon atoms, the other three signals each correspond to one carbon atom. Some of these signals will certainly be split into doublets because of the presence of carbon–phosphorus coupling. We shall also see a signal due to our solvent $CDCl_3$; this absorbs at 77 ppm and is a triplet because of coupling between carbon and deuterium (deuterium being a nucleus with spin I = 1).

The rule in carbon-13 NMR is that sp^2-hybridized carbons (carbonyl, aromatic, olefinic) absorb at lowest field, followed by sp-hybridized (acetylenic, nitrile) and sp^3 (aliphatic). A first glance leads us to believe we have seven signals, but we must remember that the methine carbon is directly bonded to phosphorus, so that we shall expect a relatively large C–P coupling. The other C–P couplings will probably be very much smaller.

So the seven signals reduce to six, one obviously being a doublet. If we expand the spectrum we see that another three signals are doublets with a small C–P coupling.

Before we try to assign the signals, let us look at the signal intensities. These are obviously not as we would expect, but are very uneven. There are two reasons for this, one having to do with the NOE and one with relaxation.

We have so far looked at the NOE only in a homonuclear manner, but of course there is also a heteronuclear NOE. Theory tells us that when we are dealing with C–H fragments in small molecules, the decoupling of the proton leads to an increase in the carbon signal intensity by up to almost 200%! So signals of protonated carbons should be stronger than those of non-protonated carbons.

Obviously we cannot however simply correlate the signal intensities with the presence of attached protons. So relaxation must also play a very important role. Relaxation times T_1 for carbon atoms also depend on whether these are protonated or not, and while T_1 for methyl or methylene groups may only be a few seconds, it may be as long as around 2 *min* for quaternary carbons! Now the choice of an ideal relaxation delay becomes impossible, and so we have to make compromizes, which result in the large variations in signal intensity.

The story is even more complicated than we have suggested, because carbon can relax by more than one mechanism. Protons rely on **dipole-dipole relaxation***, which also works well for protonated carbons but badly for non-protonated carbons. But carbon also for example makes use of* **spin-rotation** *relaxation, which is particularly active for methyl groups. And the magnetic field dependence of the various mechanisms also differs. We realize that relaxation is a very difficult subject, and if you want to know more then there are plenty of textbooks available!*

So basically there is no point in integrating a broad-band decoupled carbon spectrum. This is not so much of a drawback as it sounds, because the signals are distributed over a range of more than 200 ppm, so that line overlap is very unusual.

Signal assignment can be done in several ways: the simplest is to use prediction programmes, and Table 4 presents the result of a prediction compared with the actual values.

As we can see, the predicted chemical shifts and coupling constants agree well with the actual values.

Table 4 Result of a prediction compared with the actual values

Chemical shift (ppm)	J_{CP} (Hz)	Calculated shift	J_{CP} (calc.)	Assignment
147.1 (d)	2.3	152.0	8.0	C_{arom}–O
125.5 (s)	0	128.4	0	C_{arom}–Cl
110.9 (s)	0	115.5	4.8	C_{arom}–H
106.5 (d)	201.3	102.6	207.2	CH–P
65.2 (d)	7.2	61.9	6.0	OCH_2
16.7 (d)	5.5	15.5	8.0	CH_3

1.2.2
Coupled Spectrum (Gated Decoupling)

The proton-decoupled spectrum (Fig. 12) made it easy for us to assign the signals to the different carbon atoms, particularly because of the help given by the carbon–phosphorus coupling. However, the information which is "lost" during decoupling, the presence or absence of carbon-*proton* coupling, can be very important in many cases. Thus the degree of s-character in a C–H bond plays an important role in determining the value of $^1J_{CH}$, while the value of $^3J_{CH}$ is very important for solving stereochemical problems; the magnitude of the coupling constant $^3J_{HH}$ in an *aliphatic* fragment HC–CH was shown in the early days of NMR to depend on the **dihedral angle** subtended by the two C–H bonds, this dependence being described semi-quantitatively by the so-called **Karplus equation.** In the same way, $^3J_{CH}$ shows a Karplus-type dependence on the dihedral angle subtended by the C–H and C–C bonds involved.

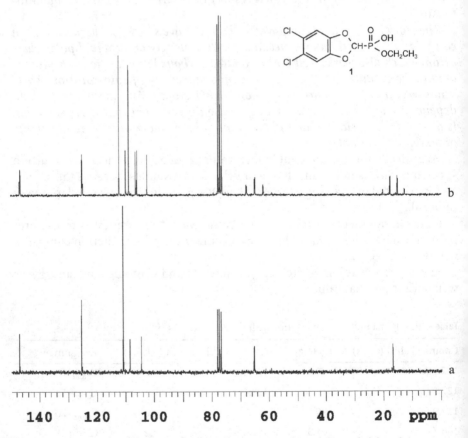

Fig. 15a,b Carbon-13 spectra of compound 1. a Protons broad-band decoupled; b carbon–proton coupling present (gated decoupling)

It is in fact quite simple to record a carbon-13 spectrum with the broad-band decoupling switched off. Such a procedure has the disadvantage that the gain in signal intensity due to the NOE is lost, so that measurement times are very long.

There is however an experiment which allows us to obtain a coupled spectrum *without* losing the NOE effect: this is known as **gated decoupling**. Here the computer has to control some elegant switching in which the broad-band decoupling is ON during the relaxation delay, allowing the NOE to build up. It is however OFF during the pulse and during the acquisition, so that we can still retain the coupling information.

Figure 15 shows the normal broad-band decoupled and gated decoupled spectra of compound 1; in the latter we can see the multiplets arising from C–H coupling (across one or more bonds) and C–P coupling. The rules for the number of lines in a multiplet and their intensities are the same as for protons, since ^{13}C and ^{31}P are both spin-½ nuclei.

1.2.3
Quantitative ^{13}C Spectrum (Inverse Gated Decoupling)

Because of the NOE and differences in relaxation rates, the intensity differences for carbon signals in a broad-band decoupled spectrum are extremely large, so that quantitative information is not available.

Though this is generally not a problem, there is an experiment available which allows us to obtain reliable quantitative intensity information, which we may for example need when studying mixtures of compounds.

This experiment is known as **inverse gated decoupling**: the broad band decoupling is OFF during the relaxation delay, so that no NOE can build up. It is however switched ON during the radio frequency pulse and during the acquisition, so that the C–H coupling is eliminated (the C–P coupling is not affected). Thus, as shown in the upper spectrum, no C–H coupling is present, and the intensities of the carbon signals are correct. The lower spectrum shows the integration values for the standard carbon-13 experiment, which are clearly completely incorrect: in each case the signal on the left is set equal to two (carbons), and while the intensities in the upper spectrum lie within 10% of the true values, most of those in the lower spectrum are too high by factors greater than two (see Fig. 16).

However, in the inverse gated experiment it is very important that the relaxation delay chosen is very long, since the carbon atoms have very different relaxation times (and relax by different mechanisms). In our example the relaxation time was set to 120 seconds! This of course makes the experiment a very time-consuming one (28 hours measurement time!).

The integration of the various carbon signals now gives intensity values which are sufficiently accurate for most purposes.

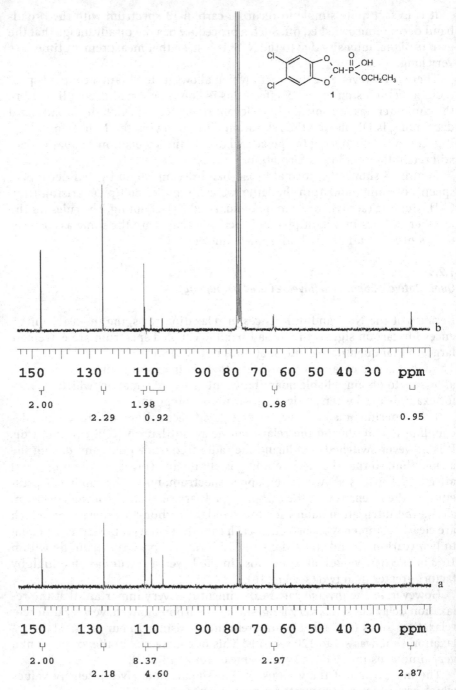

Fig. 16a,b Carbon-13 spectra of compound **1** recorded at 50 MHz. **a** Standard spectrum with integral values (measurement time 1.5 hours); **b** inverse gated decoupled spectrum with integral values (measurement time 28 hours!)

1.2.4
Decoupled Spectrum: Proton Decoupling, Proton and Phosphorus Decoupling

The signals in the coupled carbon-13 spectra are split by the C–H couplings, and the values of J_{CH} can be directly read off. If for example we consider the chlorine-bearing carbons in our model compound 1 (Fig. 17), the resulting signal is split into a doublet of doublets, due to the coupling with the two aromatic protons. The coupling paths are different: we observe both $^2J_{CCH}$ and $^3J_{CCCH}$, the values being 5.4 and 7.9 Hz respectively.

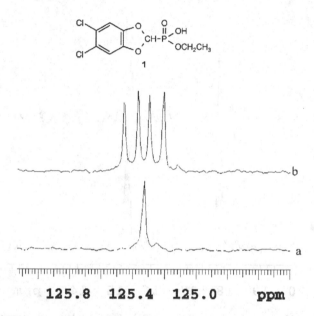

Fig. 17a,b Carbon-13 signals for the chlorine-bearing aromatic carbons in 1. a Proton decoupled; b no proton decoupling

The determination of the coupling constants is more difficult for other signals. Thus the methyl carbon of 1 (Fig. 18, lower trace) is split into a quartet by the three methyl protons. However, the four lines of the quartet are split further (into doublets of triplets), since the couplings with the P nucleus ($^3J_{POCC}$) and with the two protons of the OCH_2 group ($^2J_{HCC}$) are also readily visible.

The determination of these two coupling constants can be carried out using a **selective proton decoupling** experiment. The middle trace in Fig. 18 shows the results of such an experiment.

Here we have irradiated the OCH_2 group in the proton spectrum: the result is a doublet of quartets with two coupling constants ($^1J_{CH}$ = 127.7 Hz, $^3J_{POCC}$ 5.5 Hz). We can thus extract $^2J_{CCH}$ from the multiplets in Fig. 18b; its value is 2.7 Hz.

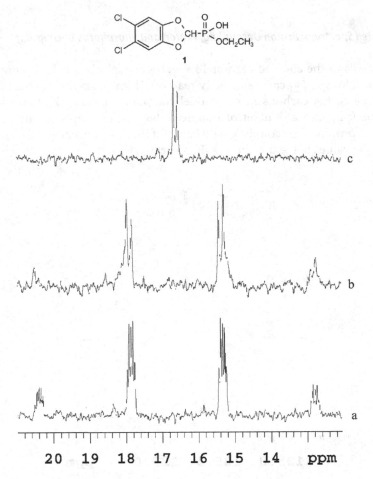

Fig. 18a–c Carbon-13 signals for the methyl carbon in **1**. a Complete carbon–proton coupling present; b selective decoupling of methylene protons; c broad-band decoupled

For completeness, the upper trace in Fig. 18 shows the broad-band-decoupled signal, which is of course just a doublet due to the P–C coupling.

1.2.5
APT, DEPT, INEPT

In the carbon-13 experiments so far discussed, only a single radio-frequency pulse has been used to irradiate the spin system. This gave us information on the chemical shifts of the carbon nuclei in the molecule. The coupled spectrum obtained using gated decoupling (1.2.2) told us how many protons are bound to any one carbon atom; however, this experiment requires a lot of time. There are however other experiments which give us this information on the "**multiplicity**" of the carbon atom (quaternary, methine, methylene, methyl) which can be carried out very quickly. Such experiments, which are invaluable in structural determination work, will be discussed here. The two most important are **APT** (Attached Proton Test) and **DEPT** (Distortionless Enhancement by Polarization Transfer).

Both find their origin in the spin-echo sequence, devised by Hahn in 1952 and used for the determination of relaxation times.

The theory behind both of these experiments, and in particular the DEPT experiment, is rather complicated, so that we refer you to NMR textbooks for details. The important feature of both is that the carbon signals appear to have been simply broad-band decoupled, but that according to the multiplicity they appear either in positive (normal) phase or in negative phase, according to their multiplicity.

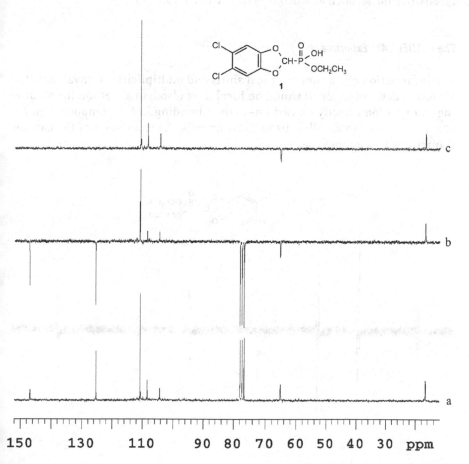

Fig. 19a–c Carbon-13 spectra of compound 1. a Standard spectrum (broad band decoupling); b APT spectrum; c DEPT-135 spectrum

APT distinguishes between two groups of signals, methyl/methine (normally shown in positive phase) and methylene/quaternary (negative). DEPT is similar, except that quaternary carbons are not detected by this sequence. There may be cases where it is necessary to distinguish between methyl and methine, and this can be done by adjusting the DEPT pulse sequence (known as "editing"): the standard experiment is known as DEPT-135, and requires, like APT, short measurement times.

Figure 19 shows the normal (broad-band decoupled), APT and DEPT-135 spectra of model compound 1. Note that in the APT spectrum the solvent ($CDCl_3$) is visible, but not in the DEPT spectrum, where the two low-field quaternary aromatic carbons are also absent.

There is another member of this family of experiments known as **INEPT** (Insensitive Nuclei Enhancement by Polarization Transfer), which was the forerunner of DEPT. INEPT still has its uses for obtaining spectra of really insensitive nuclei such as silicon-29 or nitrogen-15.

1.2.6
The INADEQUATE Experiment

The information on carbon chemical shifts and multiplicities is invaluable for structure determination. It would be ideal if we also had a method for obtaining information directly on carbon–carbon bonding in the compound under study, since this would allow us to draw on paper at least parts of the carbon framework of the molecule.

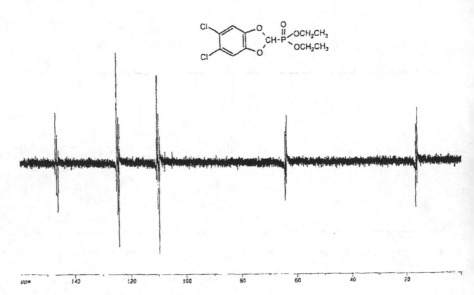

Fig. 20 1D INADEQUATE spectrum of compound 2 (75 MHz, 50% solution in $CDCl_3$, measurement time 20 hours). Note that the multiplets are distorted because they could not be correctly phased

Carbon-13 represents only 1.1% of the total carbon nuclei present in a sample. In order to get the information we require, we need to detect the doublets due to carbon–carbon coupling. Thus we wish to observe only those molecules containing two neighbouring carbon-13 nuclei, i.e. about 10^{-4} of the nuclei present; at the same time we have to get rid of the signal coming from those molecules containing only one carbon-13 (the great majority!).

INADEQUATE stands for Incredible Natural Abundance DoublE QUAntum Transfer Experiment. Again, we refer you to NMR textbooks for an explanation of the principles. Here we only present the result, which is shown in Fig. 20 for the diester 2.

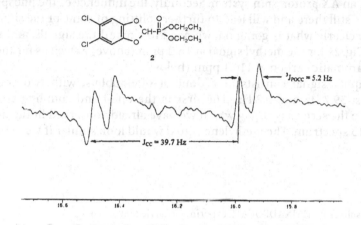

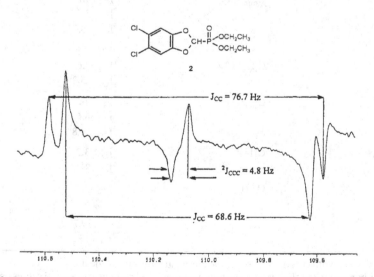

Fig. 21a,b Expansion of signals from compound 2. a Methyl carbon; b aromatic CH carbons

Note that we have used a highly concentrated solution, but even so required 20 hours to obtain the spectrum. This is because there are so few molecules containing the fragment $^{13}C-^{13}C$ present. At first glance the spectrum looks very strange, and if we count the number of signals we only find five. However, one carbon in the molecule, the methine carbon bonded to phosphorus, does not have a direct carbon neighbour, so it cannot appear. The other five signals appear at the correct chemical shifts, but they consist of multiplets which are not in phase.

Why do we see multiplets rather than singlets? Firstly, we are in each case looking at signals due to C–C coupling, so each signal will be split into a doublet just as in an **AX proton spin system**. Secondly, the influence of the phosphorus nucleus is still there and will lead to further splitting of some of the signals.

To see clearly what is going on, we need to expand the signals, and this is done in Fig. 21 for the methyl signal at 16.2 ppm (above) as well as for the protonated aromatic carbon at 110.1 ppm (below).

The upper signal consists of a doublet of doublets, with two coupling constants: 39.7 Hz and 5.2 Hz. The first is the one-bond coupling constant $^1J_{CC}$, while the second is $^3J_{POCC}$, which we have already observed in the normal carbon-13 spectrum. The methylene signal would look similar if we expanded it.

Table 5 Results from 1D INADEQUATE experiment carried out on diester **2**

Chemical shift (ppm)	Assignment	Coupling constants (Hz)
146.6	C_{arom}–O	$^1J_{CC}$ 76.7[a]
		$^2J_{COC}$ 3.2
		$^3J_{CCCC}$ 5.7[b]
		$^3J_{PCOC}$ 2.2
124.7	C_{arom}–Cl	$^1J_{CC}$ 68.6[a]
		$^3J_{CCCC}$ 5.2[b]
110.1	C_{arom}–H	$^1J_{CC}$ 76.7
		$^1J_{CC}$ 68.6
		$^3J_{CCCC}$ 4.8[b]
106.4	CH–P	$^2J_{COC}$ 3.2
		$^1J_{CP}$ 197.8
64.0	OCH_2	$^1J_{CC}$ 39.7
		$^2J_{COP}$ 6.8
16.2	CH_3	$^1J_{CC}$ 39.7
		$^3J_{CCOP}$ 5.2

[a]Should show two direct couplings, but these are apparently almost equal
[b]The two-bond coupling $^2J_{CCC}$ is smaller than the three-bond coupling $^3J_{CCCC}$ and causes only line broadening

The lower signal is more complicated, and before we can interpret it exactly we need some background information. The magnitude of one-bond C–C coupling constants depends on bond hybridization (ethane 35, ethene 68, benzene 56, ethyne 172 Hz), while two- and three-bond C–C couplings are very small, often around 2–5 Hz. The second thing we have to remember, and this is a new concept, is that the lines in the multiplets from INADEQUATE spectra often come from *different* spin systems!

Thus here we see two large doublet splittings, one between the CH carbon and the CO carbon ($^1J_{CC}$ 76.7 Hz) and one between the CH carbon and the CCl carbon ($^1J_{CC}$ 68.6 Hz).

These are due to two different $^{13}C-^{13}C$ spin systems. The third $^{13}C-^{13}C$ spin system leads to a doublet in the centre of the multiplet with a small splitting: this is $^3J_{CCCC}$ and equals 4.8 Hz.

The complete C–C and C–P coupling information is given in Table 5.

1.3
^{31}P

Phosphorus is an unusual element, because it has only one single isotope, phosphorus-31, and that this isotope is NMR-active with a spin of ½. The only other elements for which this is the case are fluorine, yttrium, rhodium and thulium.

The sensitivity of ^{31}P is also high, so that measurements do not require high sample concentrations.

1.3.1
Natural Abundance ^{31}P Spectrum of Compound 6

Since the phosphorus spectra of compounds 1 to 5 are rather boring (only one phosphorus resonance), we shall also use compound 6, which contains three non-equivalent phosphorus nuclei, to demonstrate the results of the experiments we describe.

1.3.2
Proton-Decoupled and Proton-Coupled Spectra

Compound 1 contains one phosphorus atom, so that the broad-band proton decoupled spectrum is extremely simple: it consists of only one line at 8.5 ppm. This spectrum is shown in Fig. 22, together with the proton-coupled spectrum.

The proton-coupled spectrum is much more informative. We can see immediately which protons show a measurable coupling with the phosphorus atom, because the pattern is clearly identifiable as two triplets separated by 28.7 Hz. This, as we have already seen in Table 1, is the two-bond coupling between the phosphorus and the methine proton. The triplets (intensity 1:2:1)

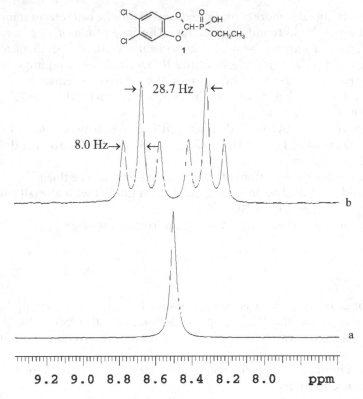

Fig. 22a,b Phosphorus-31 spectra of compound **1**. a Protons decoupled; b proton–phosphorus coupling present

result from the three-bond coupling between the phosphorus and the methylene protons, which is 8.0 Hz.

1.3.3
Coupled Spectrum (P–P Coupling)

It is by no means unusual to come across compounds which contain more than one phosphorus atom: Fig. 23 shows the proton decoupled coupled phosphorus spectrum of compound **6**, which contains three chemically different phosphorus nuclei. Phosphorus behaves in NMR just like the proton, so we shall expect to see three signals, split into multiplets if there is an observable coupling between the phosphorus nuclei.

In all cases the oxidation state of phosphorus is five, and the chemical shift range observed is only about 12 ppm. Note that the two phosphorus atoms attached to the methine carbon are non-equivalent because they are chemically different (phosphonate and phosphine oxide). We can expect the coupling between [a]P and [b]P to be large, as they are separated by two bonds, while that of [a]P to [b]P or [c]P will be small (coupling over five bonds).

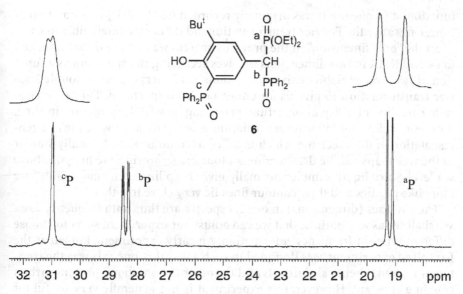

Fig. 23 Phosphorus-31 spectrum (202 MHz) of compound **6**, measurement time 2 min

We have labelled the three signals, two of which are additionally shown in an expanded form, and it is clear that the low-field signal, with the coupling of 9.4 Hz, must correspond to ᶜP. We can see only *one* other coupling (39 Hz), which occurs in both of the other multiplets: this must be between ᵃP and ᵇP. But which signal corresponds to ᵃP and which to ᵇP? This information comes from the chemical shift of analogous compounds, where phosphonates such as ᵃP absorb to high-field of 20 ppm, while phosphine oxides such as ᵇP and ᶜP absorb at around 30 ppm.

But we have a puzzle here: since rotation around the aryl-CHP₂ bond should be relatively unhindered, why does ᶜP not couple to *both* ᵃP and ᵇP? We will return to this question when we discuss the 2D phosphorus–phosphorus correlation experiment.

2
2D Experiments

2.1
General Principles, Inverse Techniques, Gradients

Two-dimensional NMR? A strange concept, when we consider that all the spectra we have previously dealt with were of course plotted in two dimensions, the two axes (dimensions) being a frequency axis (horizontal, expressed in ppm rather than in Hz for reasons we have already discussed) and an intensity axis. To understand the basic idea of two-dimensional NMR (**2D NMR**) we should first remind ourselves that while the spectrum we see and use is plotted as a function of frequency, it was originally recorded (as the FID) as a function of time. Only after the Fourier transformation did it become intelligible to us.

So the "one dimension" in the previous spectra was a time dimension, and to extend NMR to two dimensions involves recording the spectrum as a function of two time variables (time dimensions) and carrying out a **double Fourier transformation** to give us an understandable spectrum. This also naturally contains intensity information, providing us with information in three dimensions. But the intensity information is less useful, so we choose a representation of the spectrum which is called a "contour plot", basically similar to the way maps can be drawn with contour lines showing the heights above sea level. Since liquid samples normally give sharp lines, our "mountains" are more like needles and their contour lines lie very close together.

The two axes (dimensions) in our 2D spectra are thus both frequency axes. We shall see as we continue that we can adjust our experiment so as to choose different types of frequency information. An early experiment, known as the **J-resolved experiment**, was designed in such a way that one axis was the (proton or carbon) chemical shift axis and the other the one-bond proton–carbon coupling constant. However, this experiment is not generally very useful for structural determination, so that we shall not discuss it here.

The important experiments for our purposes are the *correlation experiments*, where both axes are chemical shift axes. Certainly the most useful of these is the proton–proton correlation experiment, initially known as **COSY** (for **CO**rrelated **S**pectroscop**Y**) and now, to make things more precise, as **H,H COSY**. This experiment is important, as it provides direct information on which proton nuclei couple with which.

Of course other correlations can be carried out involving any two NMR-active nuclei. The result of a **P,H correlation experiment** will be discussed below. But since most organic molecules do not contain NMR-active nuclei apart from the proton and carbon-13 (or if they do, then certainly not in 100% abundance, with the exception of fluorine-19), the other most important correlation experiments involve C and H as the relevant nuclei. These experiments, the **C–H correlation** (which can be carried out in different ways, although we shall not go into these) tell us directly which proton signal corresponds to which carbon signal. As we shall see, this type of experiment can be adjusted according to the value of the C–H coupling constant involved. We can either detect via the one-bond coupling constants or via the much smaller long-range coupling constants, and we shall see that the results are rather different.

There is also the rather famous experiment known as **2D INADEQUATE** (Incredible Natural Abundance Double QUAntum Transfer Experiment) which allows us to correlate carbon-13 with carbon-13. Potentially this experiment is very useful, since it allows us to see directly which carbon atoms are directly bonded. However, you will remember that the natural abundance of carbon-13 is only 1.1%, so a carbon-13/carbon-13 correlation requires us to detect only about 0.01% of the carbon nuclei present. Thus the experiment is very insensitive and requires large amounts of both sample and measuring time (up to 24 hours!). Since phosphorus-31 has a natural abundance of 100%, a P,C correlation experiment can be carried out much more quickly, and an example is shown below.

When we think a little more about what happens during a 2D experiment, we realize that it involves the collection and Fourier transformation of a huge amount of data. When 2D experiments were first devised, they were by no means routine. In those days computers were much slower and had much less memory. So the generation of a 2D spectrum involved several hours of measurement and quite a lot of computer time to calculate it from the raw data. Nowadays computers are much faster and have much more memory, so that 2D spectra such as H,H COSY and C–H correlation have become routine. Although we do not want to go into detail about NMR theory, we should mention that two advances in instrumentation have made 2D really fast. One is **inverse detection** (here the carbon-13 information is transferred via carbon–hydrogen coupling to the protons and the much more sensitive proton signal detected) and the other is **gradient spectroscopy** (normally we need to keep the magnetic field across the sample homogeneous, but in certain cases the application of inhomogeneous "gradient pulses", as used in medicinal NMR applications, make NMR experiments much faster). A combination of the two techniques, which is fast becoming state-of-the-art, allows us to carry out the two invaluable H,H COSY and C–H correlation experiments in minutes rather than hours! This is why we shall include them in the majority of the problems in Part 2.

In principle it is possible with many modern spectrometers to carry out correlation experiments using any two NMR-active nuclei, and we shall demonstrate this below by discussing P,C and P,P correlations.

2.2
H,H COSY

The H,H COSY spectrum of model compound 1 is shown in Fig. 24. In fact you can see a total of three spectra: the "central square" which is the actual 2D spectrum and two proton 1D spectra at the top and on the left. The computer software generates this combination of spectra automatically using a previously recorded 1D proton spectrum.

A glance at the proton spectra shows that the OH proton is missing, and when we look at the numbers along the axes we can see that in fact only the range from about 1.2 to 7.2 ppm is covered. This is a principle of 2D: only record the part of the spectrum which contains useful information! Since we want to find out which nucleus couples with which, we do not need to record the OH signal as we already know that it is a singlet.

When we discussed the 1D proton spectrum of model compound 1, we used decoupling techniques to interpret the coupling patterns. The 2D spectrum allows us to re-check our earlier conclusion. We see a singlet and three multiplets in both proton spectra. If we now look at the central square, we observe a set of four signals along a diagonal which we can draw from bottom left to top right. When interpreting an H,H COSY spectrum, the first step is to

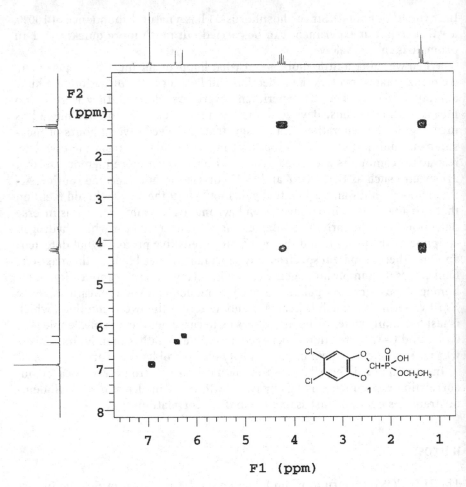

Fig. 24 H,H COSY spectrum of compound 1 in CDCl₃, measurement time 3.5 min

draw in this diagonal and identify the signals lying on it (if we are unlucky, one or more may be missing, and then we will have to adjust the height of the "contour line" using the computer).

Here all the four signals, the singlet and three multiplets, are present on the diagonal. Now we need to locate the information *on the coupling between the protons*. The doublet is caused by the proton–phosphorus coupling, you will remember, so this coupling should not be "active" in the H,H correlation spectrum. But the methyl and the methylene multiplets involve H,H coupling, which should make itself visible. It does so in the form of the "**cross peaks**" or off-diagonal peaks, and we can see two of these, one above and one below the diagonal. These are completely symmetrical with respect to the diagonal, forming a square when we draw lines to connect the signals involved. And this is the secret of interpreting H,H COSY spectra: first draw the diagonal and then locate the squares connecting the peaks on the diagonal with the off-diagonal peaks. As noted above, it is sometimes necessary to shift the contour level to make sure no squares are missing.

2.3
2D NOE

The previous experiment (COSY) demonstrated the interactions (J coupling) between protons via the bonding electrons. The NOE effect which we described in Section 1.1.6 functions because of the through-space interactions between protons, and we used the NOE difference and selective NOE experiments to demonstrate it.

NOE effects can naturally also be investigated by 2D experiments; these are known as **NOESY** and **ROESY**.

We shall use compound **3** to demonstrate the results obtained from a 2D NOESY experiment, and for comparison we shall use the COSY spectrum obtained from the same compound. Figure 25 shows the COSY (a) and 2D NOESY (b) spectra of compound **3**.

In the COSY spectrum we can see a diagonal and two sets of cross peaks: at high field those due to the coupling between CH_2 and CH_3, and at low field those due to the couplings within the aromatic ring.

Now let us look at the NOESY spectrum (b): just as in COSY, we can identify a diagonal and a series of associated off-diagonal cross peaks. Thus the interpretation of the results is analogous to the method we have already learned for COSY. However, the cross peaks are not due to spin-spin coupling but to NOE effects between the protons concerned. However, if we look more closely we can see one big difference between the diagonal peaks, which look like irregular circles, and the cross peaks, which look just like all the peaks in the COSY spectrum.

The reason for this is that our experiments are phase-sensitive. What do we mean by this? You will remember that in the DEPT and APT spectra the CH/CH_3 and CH_2 peaks are in one case positive (up) and in the other negative (down), which we also refer to as in opposite phase. Here in COSY and NOESY our experiments include such phase information, which is read off from the way the signals look in the plot.

Thus all the signals in the COSY spectrum are of the same phase, while in NOESY the diagonal and cross peaks have opposite phase.

We can see three sets of cross peaks: methyl/methylene and aromatic CH as in COSY, and in addition a clear interaction between the methine proton and the aromatic protons closest to it. This interaction is naturally not visible in the COSY spectrum, as the protons are separated by five bonds. A look back to Section 1.1.6 shows that this NOE was (as must be the case) also visible in the 1D experiment.

The advantage of the 2D NOE experiment over selective 1D NOE measurements is that all the NOEs present in a compound are detected in one single

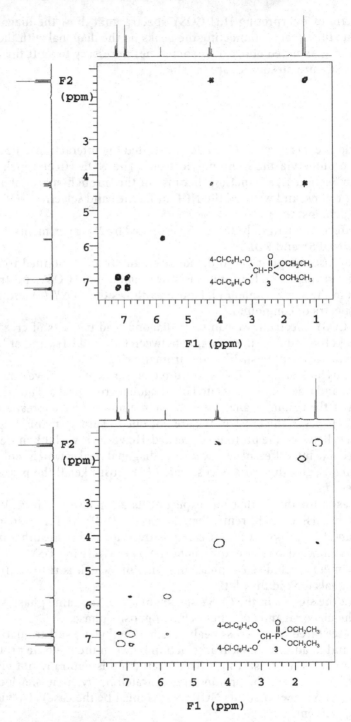

Fig. 25 2D spectra of compound 3. *Top*: COSY (200 MHz, CDCl₃, measurement time 15 min); *below*: NOESY spectrum (200 MHz, CDCl₃, measurement time 40 min)

experiment, although this takes a relatively long time. Disadvantages of 2D NOE experiments lie in the occurrence of artefacts and problems with the phase correction.

However, 2D NOE studies are invaluable in structure determination, in particular of peptides and proteins: here the NOEs give invaluable information for conformational analysis and the determination of the tertiary structures of proteins.

2.4
P,H COSY: with Varying Mixing Times for the Coupling

Since phosphorus and protons are both abundant spin-½ nuclei, it is simple to design an experiment in which we correlate protons and phosphorus rather than protons with themselves. The result of this experiment, a P,H correlation, is shown in Fig. 26. Again we have the 2D spectrum in the form of a central rectangle and two (previously recorded) 1D spectra parallel to the axes. One is the proton spectrum, the other the phosphorus spectrum. The latter of course consists of a single line, and in the 2D spectrum we do not need to look for a diagonal as there cannot be one.

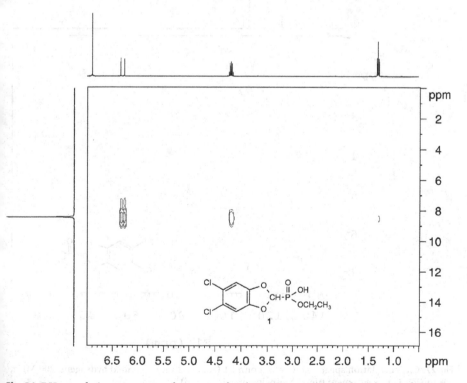

Fig. 26 P,H correlation spectrum of compound **1** (400 MHz, 5% in CDCl$_3$, delay set for J$_{PH}$ = 1.65 Hz, measurement time 12 min)

Instead, there are three rather broad contour signals corresponding to the coupling of the phosphorus with (from left to right) the methine, methylene and methyl protons. The breadth of the signals is roughly proportional to the magnitude of the coupling J involved.

This is not surprising, as the input of an average coupling constant is part of the set-up of the experiment. In fact the time period 1/J (in seconds) is involved in the experiment, and 1/J increases as J decreases. If the individual experiment is too long, the signal intensity will be decreased by relaxation.

2.5
C,H Direct Correlation

Organic compounds almost always contain carbon and hydrogen, so that the C,H correlation is a key experiment in every structural determination. This experiment tells us which carbon signal corresponds to which proton signal, and the result for model compound 1 is shown in Fig. 27.

By now we are used to the appearance of such spectra, and again the central rectangle contains the actual 2D spectrum, while the carbon spectrum (decoupled) is shown on the left and the proton spectrum at the top.

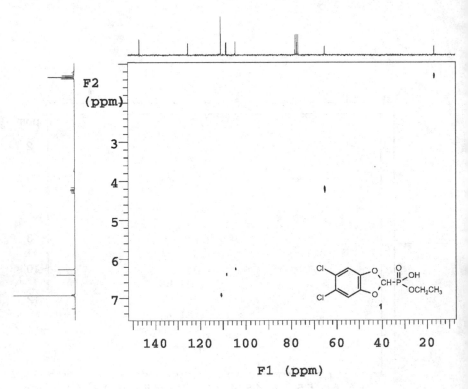

Fig. 27 C,H correlation spectrum for compound **1** (set for directly bonded hydrogens, 200 MHz, 5% in CDCl$_3$, measurement time 60 min, inverse detection)

Do not try to draw a diagonal; there is none. Probably the best thing to do when you are dealing with an unknown molecule is to construct a table, which in the present case could look like this:

H signal at	correlates with	C signal at
6.9 (aromatic)		110
6.3 (methine doublet)		106
4.2 (methylene multiplet)		65
1.3 (methyl triplet)		17

Note that, apart from the solvent signal, two aromatic carbon signals (at 125 and 147 ppm) show no correlation because they are quaternary (i.e. not bonded to protons).

Again, we need to define a coupling constant J to set up this experiment. Here for optimum sensitivity we have used an average value for direct (one-bond) carbon–hydrogen coupling constants of 160 Hz. This choice works well for most CH bonds, but is rather low if an acetylenic CH bond is present.

2.6
C,H Long Range Correlation

We can vary the result of a C,H correlation experiment by varying the coupling constant value we use (this is called varying the "mixing time"). Carbon also couples to hydrogen across two or three bonds (sometimes more), but the coupling constants are drastically lower than the one-bond coupling constant. The spectrum of 1 shown in Fig. 28 results from a long-range experiment; here J has been set to 8 Hz, which means that each *single* experiment is longer than in the direct experiment and that, due to relaxation, the signal accumulated in each experiment is smaller. However, the combination of inverse detection and gradient application makes the complete measurement fast. This long-range correlation technique is used to answer specific questions about the molecule under study. Let us construct a table for this experiment:

H signal at	correlates with	C signals at
6.9 (aromatic)		147, 125
6.3 (methine doublet)		147
4.2 (methylene multiplet)		17
1.3 (methyl triplet)		65

Because we are detecting via long-range coupling, the correlations to the methyl and methylene signals are reversed. The aromatic CH signal in the proton spectrum now correlates with both quaternary carbons, as expected. The methine doublet in the proton spectrum, however, correlates weakly with *only one* of the two low-field quaternary aromatic carbon signals; we can thus make a clear assignment of these carbons.

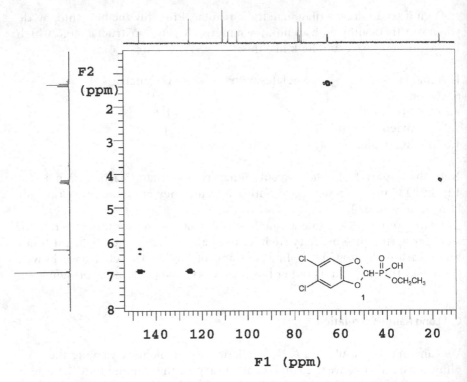

Fig. 28 C,H correlation spectrum for compound 1 (set for long-range coupling, 400 MHz, 5% in CDCl$_3$, measurement time 18 min, inverse detection)

*To be fair, we must point out that this type of experiment is extremely sensitive to the parameters chosen. Various pulse sequences are available, including the original **COLOC** (Correlation by means of Long range Coupling) as well as experiments variously referred to as **HMBC** (Heteronuclear Multiple-Bond Correlation) and **HMQC** (Heteronuclear Multiple-Quantum Correlation). Depending on the parameters chosen, it is often not possible to suppress correlations due to one-bond coupling!*

2.7
P,C Correlation

A P,C correlation experiment also requires that we use a predefined coupling constant value to determine the mixing time. A brief look at the proton decoupled carbon-13 spectrum (Fig. 14) shows that $^1J_{PC}$ is very large (around 200 Hz), while the long-range J_{PC} values are much smaller (around 5–10 Hz).

Figure 29 shows the P,C correlation for compound 1 carried out by selecting a J value of 15 Hz. The decoupled phosphorus signal is shown at the top, the proton decoupled carbon-13 spectrum on the left. The actual 2D spectrum

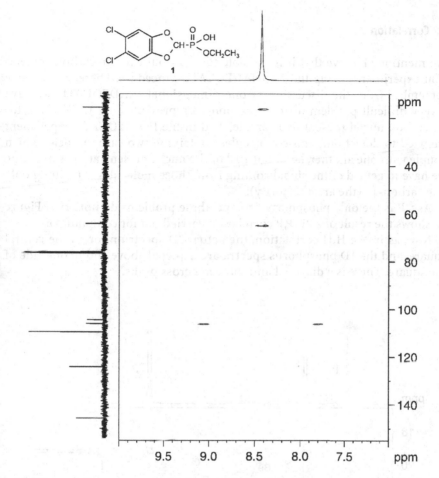

Fig. 29 P,C correlation spectrum for compound 1 (set for long-range coupling, 400 MHz, 5% in CDCl₃, inverse detection)

is in the centre and is as we might expect very simple (naturally there is no diagonal).

The first thing that we can see is that the 2D spectrum is *not* decoupled with respect to the phosphorus: the methine carbon C–P doublet in the ^{13}C spectrum is associated with a doublet along the phosphorus axis, from which $^1J_{PC}$ can of course be extracted.

Two other carbons show correlations, the methylene and methyl signals.

No correlations to aromatic carbons are visible, although the ^{13}C spectrum in Fig. 14 shows that the aromatic carbons bonded to oxygen do couple with phosphorus: if we were to carry out a second experiment using a smaller J value this correlation would however become visible.

2.8
P,P Correlation

We mentioned above that it is possible to carry out carbon–carbon correlation experiments using the 2D **INADEQUATE** procedure. There, as you may remember from the discussion of one-dimensional INADEQUATE, we have a very difficult problem to solve: carbon-13 represents only 1.1% of the total carbon nuclei present in a sample. And in the INADEQUATE experiment we need to detect only those molecules containing two carbon nuclei which couple with one another, i.e. about 10^{-4} of the nuclei present; at the same time we have to get rid of the signal coming from those molecules containing only one carbon-13 (the great majority)!

As ^{31}P is the only phosphorus isotope, these problems do not arise. Figure 30 shows the result of a 2D P,P correlation carried out for compound **6**.

Now, as in the H,H correlation, the actual 2D spectrum forms the central square, and the 1D phosphorus spectra are depicted above and to one side of the square. There is a diagonal and there are cross peaks.

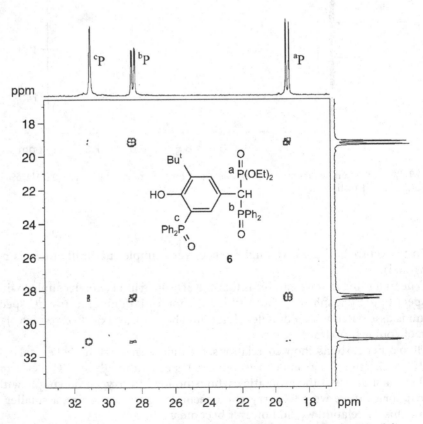

Fig. 30 P,P correlation spectrum for compound 6 (202 MHz, measurement time 15 min)

We immediately see the squares arising from the couplings between ^{3}P and ^{b}P and between ^{b}P and ^{a}P. But if we look closely we can see that there is also a weak correlation between ^{3}P and ^{a}P: this shows that there *is* a coupling between them, as we had expected. But because we can not see the coupling in the 1D spectrum the coupling constant must be smaller than the signal linewidth. This is one of the beauties of 2D correlation experiments: they often allow the detection of couplings which are not visible in the corresponding 1D spectra!

3
Quadrupolar Nucleus Experiments

3.1
General Principles: Quadrupole Moment, Relaxation, Linewidth

The experiments we have so far described have been used to study nuclei with spin I = ½ (^{1}H, ^{13}C, ^{31}P). Our model compounds 1 and 2 contain two further atoms (oxygen and chlorine), which have no NMR-active isotope with spin ½. Oxygen does however have an NMR-active isotope with spin I = 5/2 but very low natural abundance (0.037%): this is ^{17}O. Chlorine has two NMR-active isotopes: ^{35}Cl (I = 3/2, 75.53%) and ^{37}Cl (I = 3/2, 24.47%).

NMR-active nuclei with spin > ½ (these include, as we mentioned previously, deuterium) have an electric quadrupole moment and are thus referred to as **quadrupolar nuclei**.

These nuclei (and they form by far the majority of the NMR-active nuclei!) are subject to relaxation mechanisms which involve interactions with the quadrupole moment. The relaxation times T_1 and T_2 (T_2 is a second relaxation variable called the **spin-spin relaxation time**) of such nuclei are very short, so that very broad NMR lines are normally observed. The relaxation times, and the linewidths, depend on the symmetry of the electronic environment. If the charge distribution is spherically symmetrical the lines are sharp, but if it is ellipsoidal they are broad.

3.2
^{17}O

Oxygen plays a central role in organic and inorganic chemistry as well as a vital role in animal and plant life. NMR studies on this element could therefore be of great interest. Although oxygen-17 has such a low natural abundance, it is possible under correctly chosen conditions to obtain high-quality NMR spectra. Thus NMR measurements on biological materials can readily be carried out. The chemical shift range is very large (around 2500 ppm), so that in spite of the large linewidths it is possible to study structural changes readily; coupling information can also often be obtained.

Briefly, the experimental conditions should be based on the following information: acetonitrile is the recommended solvent, as it gives sharper lines than chloroform. Temperature also affects the linewidth, so that the effect of working at above room temperature should be tested. Because of the fast relaxation of the oxygen nuclei it is possible to use extremely short pulse repetition rates (50–200 msec), and the acquisition time should also be made short by appropriate choice of the number of data points and the sweep width. In this way we can record a large number of FIDs in a relatively short time. The FID needs to be subjected to exponential multiplication using linewidths of 50 to 500 Hz.

The normal reference substance is water, the signal of which is set equal to 0 ppm.

3.2.1
^{17}O Spectrum of 7: Chemical Shift (Reference), Coupling with P

We shall use our model compound 7 to show how oxygen-17 NMR can be used. Figure 31 shows the spectrum, recorded using a 40% solution of 7 in CD_3CN at a temperature of 55°C.

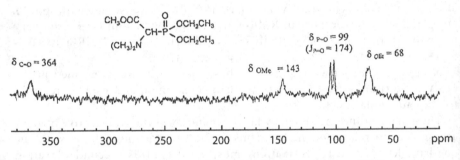

Fig. 31 Oxygen-17 spectrum for compound 7 (40% in CD_3CN, temperature 55°C)

Compound 7 contains four different oxygen nuclei, so that we expect four signals. As in carbon-13 NMR, the signal associated with the carbonyl group lies at very low field (364 ppm). The signal for the P=O oxygen at 99 ppm is immediately recognisable because of the presence of the one-bond P–O coupling (J = 174 Hz). The two remaining signals are due to the methoxy oxygen bound to carbon and the ethoxy oxygens bound to phosphorus: here the signal intensity difference indicates which is which, and the literature confirms that the high-field signal is indeed due to the ethoxy oxygens.

3.2.2
P–O Correlation

In principle it is possible (with a suitably configured spectrometer) to carry out correlation experiments between any pair of NMR-active nuclei. However, a P–O correlation is certainly not trivial, as we are dealing with a "good" (spin-½) and a "bad" (quadrupolar) nucleus. Indeed, all our attempts to carry out such an experiment failed completely.

4
HPLC-NMR Coupling

4.1
General Principles, NMR as a Highly Sensitive Analytical Tool (µg to ng Amounts)

The identification and structural characterization of biological materials, obtained for example from plants, was traditionally carried out via the classical sequence involving extraction, separation, isolation and characterization, a sequence which requires large amounts of substance and a great deal of time. Industrial problems, for example the search for small amounts of contaminants in industrial products or in waste water, also require intensive analytical studies.

A direct combination of separation and analysis techniques is thus invaluable. GLC-MS and HPLC-MS coupling are now routinely used. Because of the high sensitivity of modern NMR instruments the coupling of HPLC and NMR is now used in many NMR laboratories, and we shall discuss the principles and show some results below.

The coupling of HPLC in tandem with NMR requires two separate systems:
a) a conventional HPLC system with a standard detector (e.g. UV) and a monitoring system to observe and control the chromatography.
b) a normal NMR spectrometer with a dedicated probehead.

A long capillary with a computer-controlled switching valve (the instruments must be separated by 2–3 metres because of the strong magnetic field) connects the exit from the HPLC with the probehead. The latter is completely different in its construction from conventional probeheads: instead of the NMR tube there is a small flow cell, the volume of which is 40–100 µl. The transmitter and receiver coils are attached directly to the cell in order to maximize the sensitivity.

There are two different ways of carrying out an HPLC-NMR experiment:
a) Continuous Flow
The NMR spectrum is recorded during the chromatographic separation. Data are collected as in a 2D experiment, the two dimensions being the chemical shift and the retention time of the chromatogram.
b) Stopped Flow
Here the chromatographic scan is stopped at defined times and the NMR experiments then carried out. In this case it is possible to adjust the measurement time of the experiment to the concentration of the sample.

It is normal in conventional NMR to work with deuterated solvents, which serve both for optimising magnetic field homogeneity (lock, shim) and for avoiding the presence of the unwanted strong signals from protonated solvents.

HPLC requires much larger amounts of solvents, so that deuterated materials are too expensive; instead we work with undeuterated HPLC quality solvents, the proton signals from which are suppressed using the so-called WET sequence, which also suppress the carbon-13 satellites of the solvent signals.

Magnetic field homogeneity is ensured by the presence of D₂O at the beginning of the experiment, since many chromatographic separations use water as one solvent component. Once the homogeneity has been optimized, the coupling experiment can be carried out by changing solvent composition by replacing D₂O by pure water.

4.2
Example: Separation of 4 and 5, Two Acetals of Formylphosphonic Ester

The reaction between the cyclic orthoester **8** and diethyl chlorophosphite **9** leads via transesterification to the two acetals **4** and **5**, which cannot be separated by distillation.

4.3
Chromatogram

Figure 32 shows the HPLC chromatogram of the reaction mixture using acetonitrile and water (80:20) as solvent.

Three main peaks (1, 8 and 11) can be seen, two of which are the acetals and the third an unknown by-product.

The next step is to set the same conditions for the HPLC system which is coupled with the NMR spectrometer. The field homogeneity of the probehead is first optimized (shimmed) using the same separation column and solvent mixture.

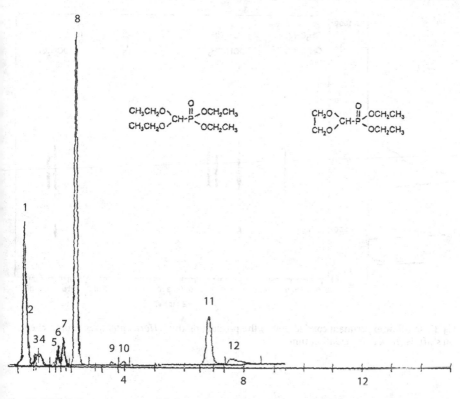

Fig. 32 HPLC chromatogram of the reaction mixture. Column: RP 18 (15 cm). Solvent: H_2O/CH_3CN (80:20). UV detector 190 nm. Pressure 157 bar. Flow rate: 1 mL/min

4.4
On-Flow Diagram (Chemical Shift vs. Time)

An on-flow experiment is now carried out. 50 µl of a solution of the product mixture (5 mg in 5 mL solvent) are injected and the NMR proton signal accumulation started simultaneously. The time taken for the chromatogram is 17 min. During this time a total of 128 proton NMR spectra are recorded, each with eight scans, i.e. an FID is accumulated approximately every 7 sec. After the Fourier transformation we obtain a two-dimensional representation (Fig. 33) of the on-flow experiment.

The two axes represent the chromatogram (retention time) and the chemical shift information. The individual NMR spectra can be extracted by the software and viewed individually in the form of normal 1D proton NMR spectra.

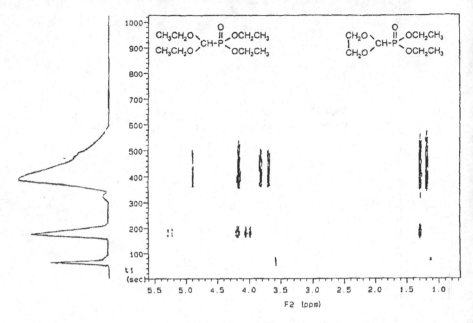

Fig. 33 On-flow experiment carried out on the product mixture. *Horizontal axis*: proton chemical shift. *Vertical axis*: retention time

The spectra recorded at retention times 1, 3 and 7.5 min are shown in Fig. 34.

The lower spectrum shows an ester group (triplet at 1.2 ppm, quartet at 4.2 ppm) and a singlet at 8.1 ppm, the latter indicating the presence of a formyl group. Peak 1 results from ethyl formate, formed by adventitious hydrolysis of the acetal 4. The middle and top spectra correspond to the two acetals (see equation); the assignment is very simple.

Acetal 5 contains only one type of methyl group, while acetal 4 has two different types. Thus the spectrum recorded after 3 min corresponds to acetal 5, while that recorded after 7.5 min results from acetal 4. The methylene groups (between 3.6 and 4.2 ppm) and the methine protons (at 4.8 and 5.4 respectively) confirm these assignments.

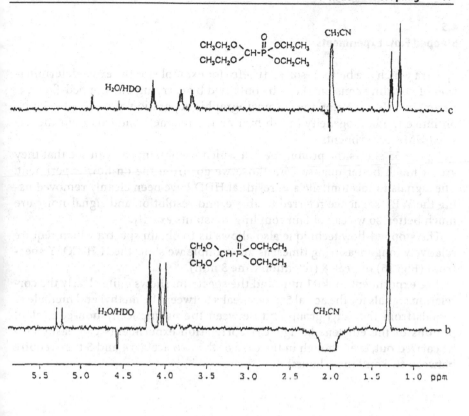

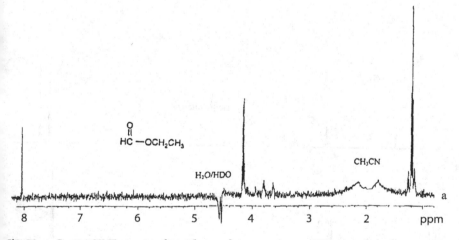

Fig. 34a–c Proton NMR spectra from the on-flow experiment. FIDs recorded after **a** 1 min, **b** 3 min, **c** 7.5 min

4.5
Stopped Flow Experiments

Spectra which are better resolved (useful for example for the exact determination of coupling constants) can be obtained by carrying out stopped-flow experiments. Here we stop the chromatographic separation after 3 and 7.5 min, optimize the homogeneity (by shimming the magnet) and carry out the desired NMR experiments.

Figure 35 shows the proton spectra which we obtain; you can see that they are of much better quality than those we got from the on-flow experiment. The signals for acetonitrile and residual HDO have been cleanly removed using the WET sequence referred to above, and resolution and signal-noise are much better, so we can obtain coupling constants exactly.

The stopped-flow technique also allows us to obtain spectra which require relatively long measuring times: as an example we show the H,H-COSY spectrum (Fig. 36) of peak 8 (retention time 3 min).

The experiment took 11 min, and the spectrum shows quite clearly the correlation signals for the acetal 5: cross peaks between the methyl and methylene signals from the ethyl group and between the magnetically non-equivalent protons of the ethylene bridge. CH correlation experiments can easily also be carried out, even though in the case of the two acetals 4 and 5 they require between two and three hours!

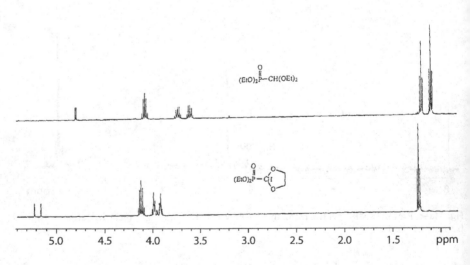

Fig. 35 Proton spectra obtained from the stopped-flow experiment. *Above*: acetal 4. *Below*: acetal 5. In each case 16 scans, relaxation delay 1 sec

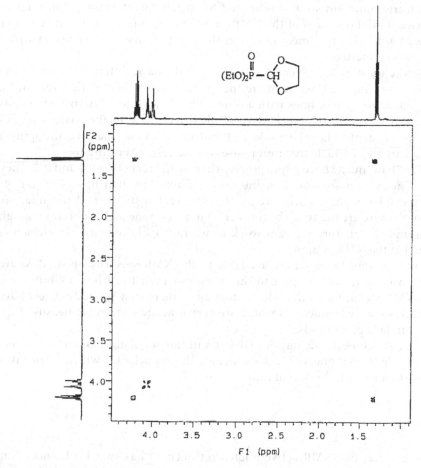

Fig. 36 COSY spectrum of acetal 5 obtained in a stopped-flow experiment. Measurement time 11 min

5
Other Spin-½ Nuclei

The series of molecules which has guided us through this book so far was chosen for a good reason: it allowed us to discuss in detail the most important nuclei, the proton and carbon-13, while demonstrating the effect of a very important "heteronucleus", phosphorus-31, on the spectra of the two key nuclei. In addition, we could discuss the NMR investigation of this heteronucleus, which exists in 100% natural abundance and has a spin of ½, and in contrast of oxygen-17, a low-abundance nucleus with a spin greater than ½.

Given the large number of elements in the periodic table, we might expect that many other nuclei are NMR-active. This is true, but the number of nuclei which are commonly studied is limited. So we felt it worthwhile at this stage to provide a brief overview of the NMR of other elements, not from the point of view of a detailed treatment but from the point of view of their use in structure determination.

Before we start, let us remind ourselves of the basic difference between the NMR-active nuclei. First there are the "good" nuclei, those with a spin of ½. These lead to narrow lines with a linewidth of the order of 1 Hz (often considerably less, not often much more). Only two of these, by the way, are single-isotope elements: phosphorus-31 and fluorine-19. As we shall see, the spin-½ nuclei are those which are of more use in structure determination.

The "bad" nuclei have a spin greater than ½, in fact between 1 and 9/2. They give lines which are broad, of the order of 100 Hz (often more but not often much less). These nuclei are much less useful in structure determination. Though there are many of them (over 50 in fact), none are important enough in terms of structure-relevant work to warrant inclusion in a text which is aimed at the NMR beginner.

Over 20 spin-½ nuclei are available to the NMR spectroscopist. Most are very insensitive with respect to the proton or even to carbon-13, but modern NMR techniques still make almost all of them easy to study. A few have NMR resonance frequencies which are very low, and cannot be measured using standard probeheads.

We have chosen nine nuclei to discuss in this section. Of course our selection is a personal one, but it does include those nuclei for which NMR gives important structure-relevant information.

5.1
^{15}N

Nitrogen has two NMR-active nuclei: nitrogen-14 has spin I = 1, and is the more abundant (99.63%). Nitrogen-15 is the spin-½ nucleus, and although it is much less abundant (0.37%) and has a low **magnetogyric ratio** (which means it is also insensitive) measurements in natural abundance are no problem with today's sensitive spectrometers. Use of techniques such as INEPT (section 1.2.6) leads to a large increase in signal strength when protons are present which couple to the nitrogen.

The total chemical shift range is close to 1000 ppm, whereby amines and nitroso compounds lie at opposite ends of this range.

IUPAC recommends CD_3NO_2 (90% in $CDCl_3$) as the chemical shift standard for both ^{14}N and ^{15}N). However, some spectroscopists reference nitrogen spectra to liquid NH_3 (not a very convenient standard!) and IUPAC recommend this as an alternative for ^{15}N. The chemical shift of CH_3NO_2 with respect to liquid ammonia is 380.5 ppm.

At this point we should mention that today spectra are often referenced to a frequency which is calculated from the deuterium reference of the solvent

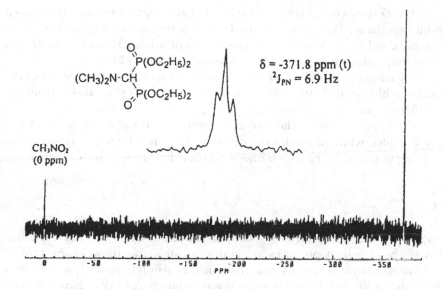

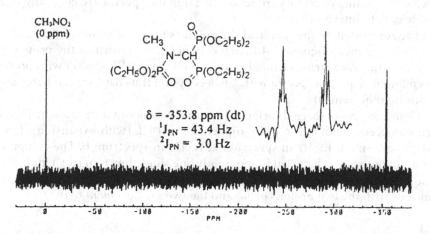

Fig. 37 Nitrogen-15 spectra of two aminophosphonates (structures as shown). 10-mm NMR tube, concentration 25% in CDCl₃, proton decoupling, relaxation delay 15 sec, measurement time 12 hours

or from the proton resonance of TMS. In the early days of NMR, when spectrometers did not have dedicated computers, it was necessary to add the standard to the sample or to put a sealed capillary containing the standard into the NMR tube containing the sample.

Figure 37 shows two nitrogen-15 spectra of aminophosphonates, referenced to nitromethane. They were recorded a few years ago, so it is interesting to note the conditions: 10-mm NMR tube, concentration 25% in CDCl₃, proton decoupling, relaxation delay 15 sec, measurement time 12 hours (!).

Today we might be able to carry out the measurement a little faster, but the results would not be better. The lines are very sharp, as you can see from the fact that we can clearly determine coupling constants as low as 3 Hz.

In the upper spectrum the nitrogen couples to two phosphorus nuclei to give a triplet, while in the lower spectrum there is a large (43.4) one-bond coupling to give a doublet, each line of which is in turn split into a triplet.

5.2
¹⁹F

Fluorine-19, like phosphorus-31, is a spin-½ nucleus with 100% natural abundance. The signals it produces are almost as strong as those of the proton, and the resonance frequency at a given field is also relatively close to that of the proton. Although for many years it was in fact necessary to have a special probehead for fluorine-19, those days have gone and fluorine has become a completely "normal" nucleus.

The total chemical shift range is over 1000 ppm, so that although fluorine-element coupling constants are relatively large the spectra are generally relatively easy to interpret.

The zero-point on the chemical shift scale is the δ-value for CFCl₃.

The resonance frequency of fluorine-19 lies close to that of the proton, so that the same measuring channel is used to observe it. ¹⁹F spectra with proton decoupling or proton spectra with ¹⁹F decoupling thus have special hard- and software requirements.

Figure 38 shows three fluorine-19 spectra: a potassium fluoride in D₂O; b trifluoroacetic acid; and c p-fluorophenol in CDCl₃ (with expansion). Linewidths are small: 1.9 Hz in spectrum a, 1.3 Hz in spectrum b. The computer printout in c shows that what is apparently one single line is in fact a multiplet, and the expansion shows a complex multiplet due to coupling of the fluorine nucleus with the two protons *ortho* and the two protons *meta* to it.

5.3
²⁹Si

Silicon is one of the most abundant elements in the earth's crust. We find silicon in sand and quartz, and in our NMR tubes. Of course we also find it in the computers which run our NMR spectrometers.

As far as NMR is concerned, the spin-½ nucleus silicon-29 has a natural abundance of 4.7%. The chemical shift range is around 600 ppm, and the shift of TMS is the zero-point.

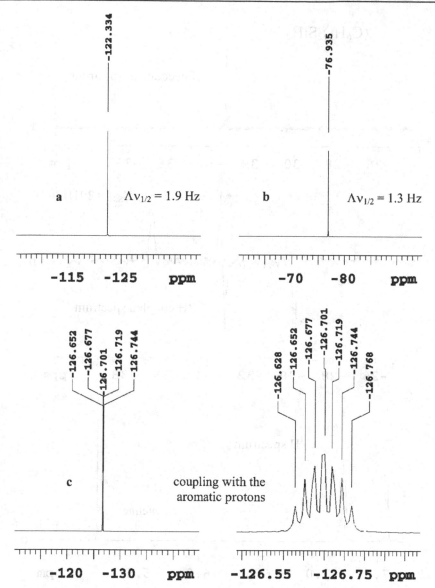

Fig. 38a–c Fluorine-19 spectra: **a** potassium fluoride in D$_2$O; **b** trifluoroacetic acid; and **c** p-fluorophenol in CDCl$_3$ (with expansion)

To a large extent the chemical shifts of carbon and silicon run parallel, but the chemistry of the two elements is somewhat different. Thus silicon can have extend its valence shell beyond the coordination number of 4. A few stable organosilicon compounds in which silicon is divalent are known (the silylenes), and compounds with a silicon–silicon double bond also exist (the disilenes).

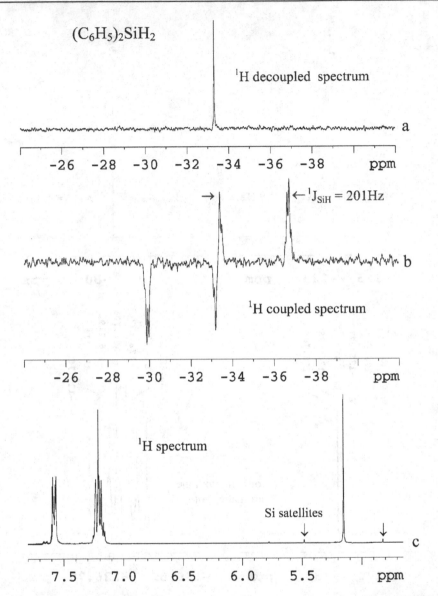

$(C_6H_5)_2SiH_2$

^1H decoupled spectrum

a

-26 -28 -30 -32 -34 -36 -38   ppm

$\leftarrow ^1J_{SiH} = 201Hz$

b

^1H coupled spectrum

-26 -28 -30 -32 -34 -36 -38   ppm

^1H spectrum

Si satellites

c

7.5   7.0   6.5   6.0   5.5   ppm

Fig. 39a–c Silicon-29 and proton spectra of diphenylsilane in C_6D_6. **a** INEPT spectrum with complete proton decoupling, **b** proton-coupled INEPT spectrum ($^1J_{SiH}$ 201 Hz); the fine structure is due to coupling with thearomatic protons, **c** proton spectrum showing ^{29}Si satellites for the SiH protons)

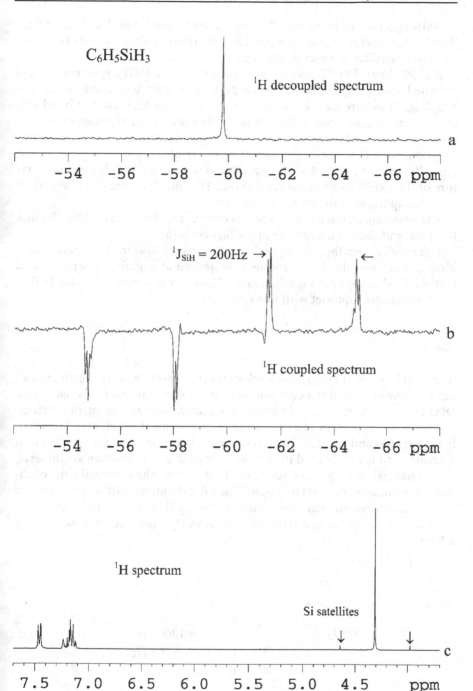

Fig. 40a–c Silicon-29 and proton spectra of phenylsilane PhSiH₃ in C₆D₆. **a** INEPT spectrum with complete proton decoupling, **b** proton-coupled INEPT spectrum ($^1J_{SiH}$ 200 Hz); the fine structure is due to coupling with the aromatic protons, **c** proton spectrum showing ^{29}Si satellites for the SiH protons)

NMR spectra can be recorded using various techniques. INEPT (see Section 1.2.5) is useful, because the **polarization transfer** from protons to silicon leads to a considerable increase in signal intensity.

Fig. 39 shows INEPT spectra of diphenylsilane PhSiH$_2$: spectrum **a** was recorded with complete proton decoupling, spectrum **b** without proton decoupling. Thus here we see a triplet structure from which the one-bond silicon–proton coupling can be determined. The value of 201 Hz means that care must be taken in the proton decoupling experiment to remove such a large coupling. Note that the central signal of the triplet is partially positive and partially negative, while the left-hand signal is negative. This is due to the nature of the INEPT experiment carried out. The fine structure in the signals is due to coupling with the aromatic protons.

The lower spectrum **c** is a proton spectrum. The SiH protons absorb close to 5 ppm and show satellites due to the SiH coupling.

Figure 40 shows the corresponding set of three spectra for phenylsilane PhSiH$_3$. Note that the Si–H protons now absorb at slightly higher field and that the Si–H coupling is slightly smaller. Naturally the proton coupled INEPT spectrum shows a quartet with fine structure.

5.4
^{77}Se

Although it is toxic in large doses, selenium is an essential micronutrient in all known forms of life. It is a component of the unusual amino acids selenocysteine and selenomethionine. In humans, selenium is a trace element nutrient.

The most important use of selenium is as a pigment which gives a red colour to glasses and enamels. However, selenium is a catalyst in many chemical reactions and is widely used in various industrial and laboratory syntheses.

The natural abundance of selenium-77 is 7.58%. The chemical shift of dimethyl selenide is set equal to 0 ppm. The total chemical shift range is around 2200 ppm, organoselenium compounds covering almost the whole range.

Figure 41 shows the spectrum of H$_2$SeO$_3$ in D$_2$O, the linewidth being only 2.5 Hz.

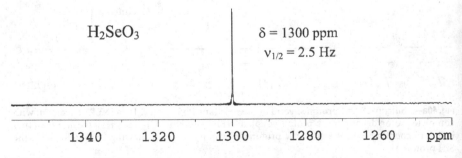

Fig. 41 Spectrum of H$_2$SeO$_3$ in D$_2$O

5.5
^{113}Cd

There are two main uses for cadmium: in batteries (particularly Ni–Cd batteries), which account for almost three quarters of the consumption, and in pigments and plastics stabilizers.

Cadmium and its compounds are highly toxic and can be carcinogenic. Thus great care should be taken when working with this element!!

There are two magnetically active isotopes of cadmium, cadmium-111 and cadmium-113, with natural abundances of 12.80 and 12.26% respectively; the latter is studied, as the signal is slightly stronger. The chemical shift of $Cd(ClO_4)_2$ (0.1M) in D_2O is defined as zero ppm, the secondary standard being dimethyl cadmium (642.93 ppm). The total chemical shift range is around 900 ppm, that of organocadmium compounds ca. 400 ppm.

5.6
^{117}Sn, ^{119}Sn

Tin has many uses, including: coating (tins/cans for food), alloys such as bronze, organ pipes, solder, and the float glass process. It is also important in laboratory syntheses, in spite of the well-known toxicity problems.

Tin is an unusual element in that it has three magnetically active isotopes, all spin-$\frac{1}{2}$. However, tin-115 has a natural abundance of only 0.35%, and is never studied. The other two, tin-117 and tin-119, occur in similar amounts (7.61 and 8.58% respectively). Spectra of the latter are normally recorded, as it is about 25% stronger. Tetramethyltin is taken as the zero-point, and the total chemical shift range is about 3000 ppm.

Tin chemical shifts run broadly parallel to those of silicon; tin can readily increase its coordination number to five or six. A few stable organostannylenes R_2Sn and distannenes $R_2Sn=SnR_2$ are known.

Thus tin chemical shifts are of considerable use in structural studies. In addition, tin-element coupling constants are easily visible, and particularly in proton and carbon-13 spectra the relevant coupling constant values are of diagnostic use: for example both tin–proton and tin–carbon coupling constants show a **Karplus**-type behaviour.

Vinyltin compounds are very important in organic synthesis, since the vinyl moiety can be readily transferred to carbon in the (palladium-catalyzed) Stille reaction. The transfer is stereospecific, and the geometry of the vinyltin moiety can be easily checked using proton and carbon-13 NMR via the coupling satellites.

Figure 42 shows spectra of the simplest tetraorganotin compound, tetramethyltin. The upper spectrum was recorded with complete decoupling of all protons, the middle spectrum without. The result is a multiplet with 13 lines (n = 12), but if you work out the binomial coefficients for such a multiplet you will see that the outer two lines are too weak to be seen. The lower spectrum is the proton spectrum, which shows satellites due to two-bond tin–proton coupling to the tin-117 (inner lines) and tin-119 (outer lines) nuclei.

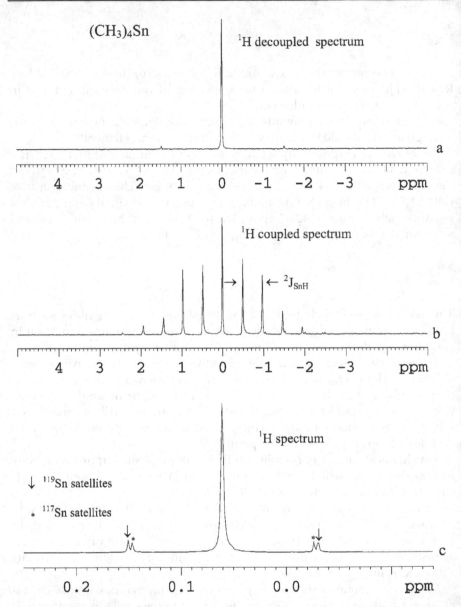

Fig. 42a–c Spectra of tetramethyltin in CDCl$_3$. **a** Proton decoupled, **b** proton coupled (^{2}J$_{SnCH}$ 54.3 Hz), **c** proton spectrum. The satellite signals are due to coupling to tin-117 (*inner lines*) and tin-119 (*outer lines*). The ratio of the coupling with tin-119 to that with tin-117 is 1.046:1 (the ratio of the **magnetogyric ratios** of the two nuclei)

The tin spectrum becomes much more interesting when the molecule under study contains two tin nuclei which can couple with each other. Figure 43 shows the spectrum of a 1,1-distannyl-1-alkene. The tin nuclei are separated by two bonds, so that a large tin coupling can be observed. The signals

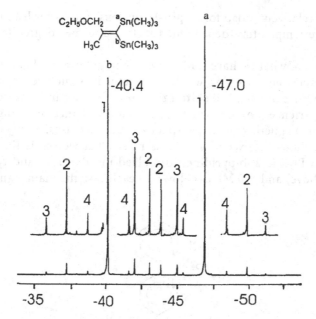

Fig. 43 Tin-119 spectrum of a 1,1-distannyl-1-alkene (structure shown). Signals result from various isotopomers: 1 from molecules containing one tin-119 nucleus, 2 from molecules containing one tin-119 and one tin-117 nucleus, 3 from molecules containing two tin-119 nuclei and 4 from molecules containing tin-119 and carbon-13 nuclei

marked 1 in the figure are due to molecules which contain only one tin-119 nucleus. A number of other isotopomers (isomers containing different isotopes) are also visible, and can be seen better in the upper trace. Molecules containing one tin-119 and one tin-117 nucleus are marked 2 and form an AX spin system; thus they are symmetrical with respect to the lines 1. Molecules containing two tin-119 nuclei lead to the signals marked 3, and form an AB spin system. The coupling constant $^2J(^{119}Sn-^{119}Sn)$ is 684 Hz.

Finally, molecules containing tin-119 and carbon-13 isotopes (in the methyl groups bound to tin) lead to the small signals marked 4.

5.7
^{195}Pt

The main uses for platinum are as a catalyst in the catalytic converter and in fuel cells. And of course platinum, a very expensive metal, is used in jewellery. However, certain platinum-containing compounds are chemotherapeutic agents, examples being cisplatin, carboplatin and oxaliplatin. This explains the synthetic interest in platinum compounds.

Platinum-195 is the only magnetically active isotope of platinum, the natural abundance being 33.8%. The shift of a saturated solution of K_2PtCl_6 is in D_2O defined as zero ppm. The total chemical shift range is huge, about 13,000 ppm (from –6000 to +7000 ppm!).

Lines are relatively broad for a spin-½ nucleus, and as for lead the chemical shifts are temperature-dependent (up to 1 ppm per degree temperature change).

In the light of what we have said above, we might expect that satellites due to platinum-element coupling would be useful in structure determination. However, because of **chemical shift anisotropy** they are in fact often not visible, and experience (and theory) suggest that the chance of seeing them decreases as the magnetic field of the spectrometer increases.

Figure 44 shows an exception to this rule. What we see is the signal for protons 6 in PtCl$_4$(2,2'-bipyridine), dissolved in DMSO-d6 and recorded at 200 MHz (above) and 600 MHz (below). In each case the main signal consists

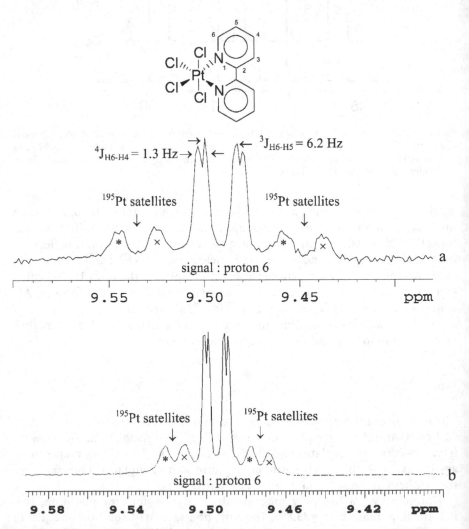

Fig. 44a,b The signal for protons 6 in PtCl$_4$(2,2'-bipyridine), dissolved in DMSO-d6 and recorded at 200 MHz (a) and 600 MHz (b)

of a doublet of doublets resulting from coupling between protons 6 and protons 4 and 5 (the values of the proton–proton coupling constants are given). And in both cases the satellites due to coupling with platinum-195 are clearly visible ($^3J_{PtH}$ 26 Hz).

Let us turn to the platinum spectra themselves. Figure 45 consists of three spectra. The top spectrum a shows the platinum signal from an inorganic salt, K_2PtCl_4. Note the linewidth of 45 Hz, relatively small in comparison with the

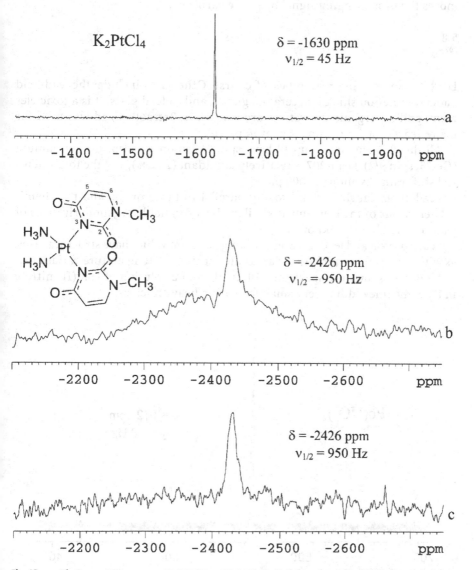

Fig. 45a–c Platinum-195 spectra, 64.52 MHz. **a** K_2PtCl_4 in D_2O, **b** and **c** *cis*[Pt(NH₃)₂(1-methyluracil-N3)]. Spectrum **b** was recorded using a normal pulse sequence, with 90° pulses. Spectra **a** and **c** were recorded using the ARING pulse sequence for removing acoustic ringing

other two spectra. These, **b** and **c**, show signals from *cis*[Pt(NH$_3$)$_2$(1-methyl-uracil-N3)]. The spectra are identical except for the presence of a very broad signal in spectrum **b**, which is due to a probehead phenomenon called **acoustic ringing**.

This effect is caused by an RF pulse through a conductor such as the NMR coil in a magnetic field Acoustic ringing increases at lower NMR frequencies and special pulse sequences are used to suppress it.

Spectrum **c** was recorded using a pulse sequence called ARING, which removes the broad ringing signal by phase variation.

5.8
^{207}Pb

Lead is also used in organ pipes, of course. Other uses include: the lead-acid battery, radiation shielding, ceramic glazes, and in lead glass. It is a toxic element, and its organic derivatives are also toxic. Tetraethyllead was used for many years as an anti-knock agent in petrol.

All the elements in Group 14 have spin-½ isotopes except for germanium (Ge-73, spin 9/2). Lead-207 is relatively abundant (22.6%), and the total chemical shift range is about 20,000 ppm.

Lead chemical shifts run broadly parallel to those of tin, and the chemical behaviour of the elements is similar. However, stable organoderivatives of lead (II) are almost unknown.

Lead–proton and lead–carbon coupling constant values have structural uses, as with tin. Lead chemical shifts are quite sensitive to temperature variations.

Figure 46 shows the spectrum obtained from a solution of lead(II) nitrate in D$_2$O; the linewidth is very small for such a heavy nucleus.

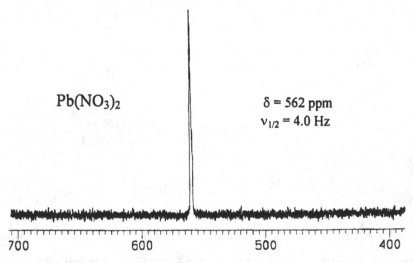

Pb(NO$_3$)$_2$

δ = 562 ppm
$\nu_{1/2}$ = 4.0 Hz

700   600   500   400

Fig. 46 Lead-207 spectrum of lead(II) nitrate in D$_2$O

6
Solid State NMR

6.1
General Principles

In Section 1.1.1 of this book, we said that "The proton spectra are normally measured in 5-mm sample tubes, and the concentration of the solution should not be too high to avoid line broadening due to viscosity effects". Taking this statement to its logical conclusion, we might well expect that NMR measurements on solids would be impossible. Indeed, if we part-fill a 5-mm glass sample tube with a proton-containing solid, put it into the spectrometer and rotate it at around 20 Hz, we shall not detect any proton signal.

The reason for this is that a series of strong interactions within a solid sample broaden the linewidth to such an effect that no signal appears to be present under these measurement conditions.

These interactions are
- *the Zeeman interaction with the external magnetic field (i.e. the normal splitting of energy levels)*
- *the magnetic shielding by the electrons (to give the chemical shift)*
- *spin-spin coupling to other nuclei.*

Naturally these three are also present in solution, but in addition in solids there are:
- *direct dipole-dipole interactions with other nuclei, which depend on the magnitude of the nuclear magnetic moments and are most important for spin-½ nuclei with large magnetic moments such as H, F and P. They are independent of the external field, and dependent on the internuclear distance*
- *finally, for quadrupolar nuclei there is the field-independent quadrupolar interaction, which is normally the dominant effect in the spectra.*

Fast motions of the molecules in the liquid state average all these interactions. Chemical shifts and J values are measured as discrete averages, and the dipolar and quadrupolar interactions are averaged to zero.

Averaging does not occur in the solid state, so that spectra are normally more complex, but also contain more information.

The use of a combination of two techniques can however remove or decrease these interactions to such an extent that NMR spectroscopy of solid samples becomes possible.

These two techniques are called **cross-polarization (CP)** and **magic angle spinning (MAS)**: in combination, these are thus called **CP-MAS**.

CP-MAS-NMR of solids is now almost a routine technique, but unfortunately there are few modern textbooks which deal with solid state NMR, so that we felt it important that you should at this stage learn something about this area of NMR, so that you know of its existence and something about its possibilities.

We shall not go into the details of cross-polarization, as they are much too complex for the beginner. Magic-angle spinning is much easier to deal with: in the mathematical expressions for the internuclear interactions there is a factor $(3\cos^2-1)$. This reduces to zero when the angle is 54°44'. This angle is called the "magic angle", and it refers to the axis around which the sample is spun (rotated) relative to the (vertical) axis of the NMR spectrometer. Spinning at 20 Hz is however not enough, and in order to remove (at least partially) another effect called **chemical shift anisotropy** (see below) the sample needs to be spun at rates between 5 and 35 kHz!

Since spins and the external magnetic field are all vectors (i.e. they have both magnitude and direction), interactions between them must be described by a 3x3 matrix or "tensor" which characterizes the interaction.

Thus the chemical shift in the solid state has three components in the directions of 3 orthogonal axes. MAS allows us to obtain the isotropic chemical shift, the quantity which we measure in solution and which thus interests us.

We can imagine that a normal glass sample tube is not suitable for this kind of treatment, and the sample (which should be amorphous or consist of powdered crystals) is packed tightly into a so-called rotor, generally ceramic in nature (ZrO, is often used) and 4 mm in diameter (and rather expensive!). This rotor is placed in a special solids probehead, and the magic angle is set automatically. Because of the nature of the interactions referred to above, it is best to use spectrometers with a proton resonance frequency of 300–400 MHz; higher fields bring disadvantages rather than advantages.

The quantity of sample required lies between 10 and 100 mg on average, and the sample is not destroyed, but can be recovered completely.

6.2
Solid State ^{1}H NMR

The combination of cross polarization (basically a pulse sequence) and MAS is sufficient to drastically reduce the linewidths of spin-½ nuclei. Liquid-state proton NMR spectra, as we have seen, are characterized by extremely narrow lines and complex multiplets due to spin-spin coupling; in addition, the normal chemical shift range is only around 10 ppm.

No solid state NMR experiment is able to obtain spectra comparable to those routinely recorded in the liquid state. Thus multiplets become broad singlets and, if close together, overlap to give a broad signal "envelope". While the differentiation of aromatic and aliphatic protons is simple, the information available is, from the point of view of structure determination, very limited. Thus we shall not provide an example.

6.3
Solid State ^{13}C NMR

The situation for carbon-13 is however completely different. Here spectra are normally recorded with complete proton decoupling, so that we see a series of single lines; in addition the chemical shift range is around 200 ppm. Magnetization transfer from protons to carbon-13 leads to an increase of signal intensity by a factor of up to 4. The pulse repetition rate can be increased when CP is used, because of the fast relaxation of the protons. Thus the slight line broadening characteristic of solid state spectra is no problem. This line broadening is particularly small if the molecule in question crystallizes to give so-called "plastic crystals", a well-known example being camphor.

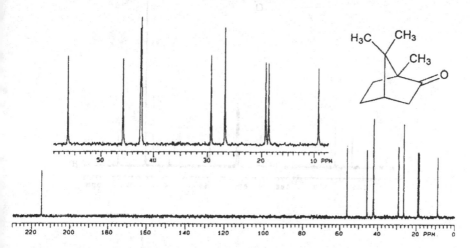

Fig. 47 Carbon-13 solid state NMR spectrum of camphor (50.309 MHz, cross-polarization contact time 5 ms, spin rate 1650 Hz)

Generally, however, linewidths in solid state spectra are grater than in solution. An example is shown in Fig. 48, which compares the solution spectrum (a above) of a bisphosphonate (the structure of which is shown in the Figure) with that in the solid state (b below). While the lines are much sharper in the upper spectrum, we can clearly see splittings in the solid state spectrum. Since the protons are decoupled, these must be due not to coupling but to non-equivalence. In the solid state the molecule becomes unsymmetrical due to hydrogen bonding of the type (–OH---O=P), so that the carbon nuclei (e.g. in the t-butyl groups) give separate signals. Solid state NMR spectra make different spatial arrangements of groups in a molecule visible.

This is where solid state NMR starts to come into its own. Molecules which form crystals can be studied routinely by X-ray crystallography, which gives exact structural information. This information is limited in value, however, since synthetic chemistry is generally carried out in the liquid and not the solid phase. If the crystals are powdered and subjected to solid state NMR, we obtain information which can (not always, but often) be compared with that obtained when the same substance is studied by solution NMR.

Thus solid state NMR can be considered as a (potential) "bridge" between solution NMR and X-ray crystallography.

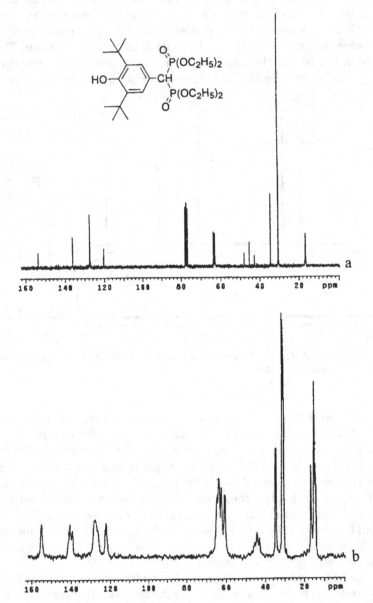

Fig. 48a,b Carbon-13 spectra of a bisphosphonate (structure given) in solution (a, in CDCl₃) and in the solid state (b, CP/MAS, spinning rate 12 kHz)

6.4
Solid-State ^{31}P NMR

We have referred to the various interactions which can cause line broadening in the solid state. One of these, which is normally not a problem in liquid state NMR, is due to the fact that the chemical shift itself is a **tensor**, i.e. in a coordinate system with orthogonal axes x, y and z its values along these axes can be very different. This **anisotropy** of the chemical shift is proportional to the magnetic field of the spectrometer (one reason why ultra-high field spectrometers are not so useful), and can lead in solid state spectra to the presence of a series of **spinning sidebands**, as shown in the spectra of solid **polycrystalline** powdered triphenylphosphine which follows (Fig. 49). In the absence of spinning, the linewidth of this sample would be around 75 ppm!!

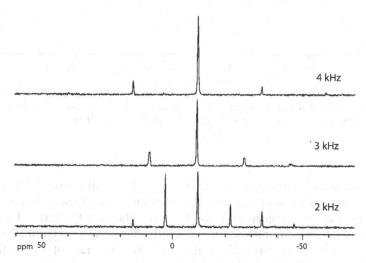

Fig. 49 Phosphorus-31 CP/MAS spectra of polycrystalline triphenylphosphine (powder) at the given spinning rates

By raising the spinning rate from 2 to 4 kHz the intensity of these sidebands (the distance from the central signal to the sidebands is equal to the spinning rate) decreases vastly, and at even higher spinning speeds no sidebands occur. The position of the central signal is independent of the rate of rotation. Note that in the spectrum measured at 2 kHz the spinning sidebands to the left and the right of the central signal are not symmetrical in their intensity. This is typical, and in some spectra measured at low spinning rates the central signal, i.e. the "real" (**isotropic**) chemical shift, corresponding (but not necessarily equal) to the chemical shift we measure in solution NMR, is not even the strongest!!

Many catalysts, both immobilized (on solid state supports) and heterogeneous, contain phosphines and other phosphorus compounds, so that solid state NMR has become an invaluable tool in the study of catalysis.

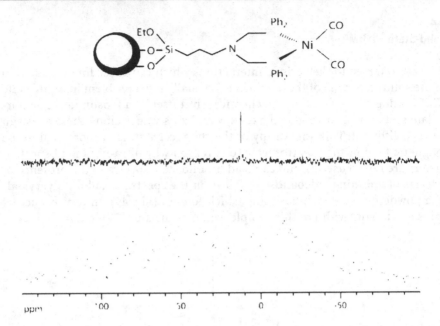

Fig. 50 Phosphorus-31 CP/MAS spectrum of the nickel complex shown, which is immobilized on silica gel, recorded at a spinning rate of 4 kHz (*lower spectrum*) and HRMAS spectrum of an acetone suspension, recorded at a spinning rate of 2 kHz (*upper spectrum*)

If the samples are amorphous rather than polycrystalline (as is the case in catalysts attached to solid surfaces) the linewidths are greater. This is shown in our next example: Figure 50 shows two phosphorus-31 spectra of a nickel complex which is immobilized on silica gel. The lower spectrum is a "conventional" CP/MAS spectrum. In spite of the rotation rate of 4 kHz, the increase in **chemical shift anisotropy** leads to the presence of a number of spinning side bands. The upper spectrum shows a so-called HRMAS spectrum recorded using a suspension in acetone at a spinning rate of 2 kHz. Although the signal-to-noise ratio is poor, a sharp single line is observed. Commercially available HRMAS rotors (HR = high resolution) are suitable for such suspension measurements. Unfortunately not all supported materials are suited to suspension measurements.

Measurements made on a group of 3 structurally similar organophosphorus compounds containing one, two and three phosphorus nuclei shows the power of solid state NMR. The relevant data are given in Table 6, while Fig. 51 shows the three MAS spectra.

The solution spectra of A–C show, as expected, only one signal in the phosphorus-31 spectra. However, the MAS spectra show either one, two or three signals, reflecting the number of phosphorus nuclei present. The X-ray crystal structures of A–C show that there is hydrogen bonding between the OH group and the phosphoryl oxygen(s). Depending on the number of the P=O groups in the molecule the hydrogen bonding is different, so that the two or three phosphorus atoms are non-equivalent in the crystal. This non-equivalence makes itself felt in the MAS spectra.

Table 6 P high resolution and MAS spectra of three phosphonic acid esters (A–C)

Compound		δ (ppm) in CDCl	δ (ppm) MAS
HO $\times$ CH₂ P(OC₂H₅)₂ O	A	27.3	23.8
HO CH P(OC₂H₅)₂ / P(OC₂H₅)₂ O	B	19.2	17.3 18.7
HO C P(OX OC₂H₅)₂ P(OC₂H₅)₂ O	C	16.8	14.1 16.5 17.9

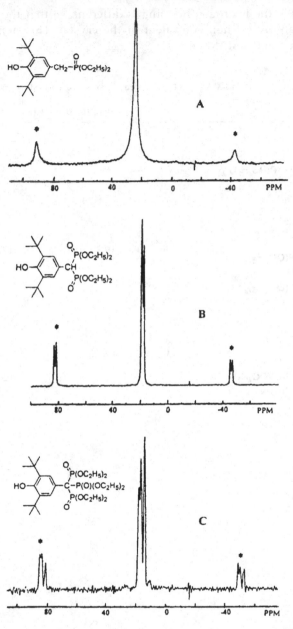

Fig. 51A–C Phosphorus-31 MAS spectra of compounds **A–C**. Signals marked with an *asterisk* are due to spinning sidebands

6.5
Solid-State ^{29}Si NMR

In CP/MAS spectra of silicon the silicon signals are magnified by a factor of up to 5. We will return to catalysis to give an example of another advantage of CP: silica gel can be surface-modified in various ways, and CP/MAS allows us to study the modified centres without disturbance by the bulk species. Figure 52 shows the silicon-29 CP/MAS spectrum of silica gel modified by the introduction of a phosphine linker. We can see a total of five signals, that at highest field (ca. –110 ppm) being due to unmodified SiO$_4$ centres, the other four to the structural fragments shown (clockwise).

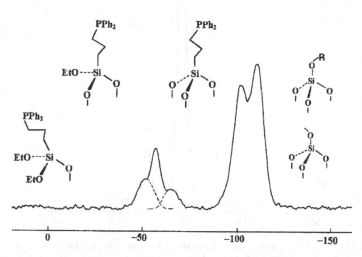

Fig. 52 Silicon-29 CP/MAS spectrum of silica gel modified by the introduction of a phosphine linker. The signal at highest field (ca. –110 ppm) is due to unmodified SiO$_4$ centres, the other four to the structural fragments shown (*clockwise*)

6.6
Solid State NMR

The chemical shift anisotropy of tin is large, so that many spinning sidebands are often seen at low spinning speeds. Linewidths can vary greatly, but species where tin is bound to four carbon atoms often give narrow lines and small spinning sidebands which can readily be removed by spinning at higher rates.

It was thus of interest to see whether satellites due to tin–tin coupling could be observed. The first case where this was possible was tetrakis(trimethylstannyl)methane, the spectrum of which is shown in Fig. 53.

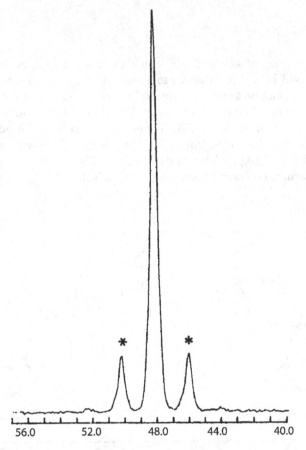

Fig. 53 74.63 MHz CP-MAS spectrum of $C(SnMe_3)_4$, isotropic chemical shift 48.2 ppm, $^2J(^{119}Sn-^{119}Sn)$ 328 Hz (the coupling visible is that between a tin-119 and a tin-117 nucleus). Both the isotropic chemical shift and the two-bond tin–tin coupling correspond to the solution values

A series of tetrastannylcyclohexanes was also studied by CP/MAS. Figure 54 shows the central part of the spectrum of one of them, the formula of which is shown. An X-ray crystal structure showed that there are four non-equivalent tin atoms in the unit cell, so that the presence of four lines in the CP/MAS spectrum was not unexpected. Satellites due to one-bond tin–tin coupling are clearly visible. This coupling is very large, over 4000 Hz, and in fact there are two pairs of lines with very slightly different couplings (lines 1 and 3 correspond to a coupling of 4162 Hz, lines 2 and 4 to a coupling of 4240 Hz (the solution value is 4245 Hz). The difference is 1.9%, compared with a measured bond length difference of 0.04%.

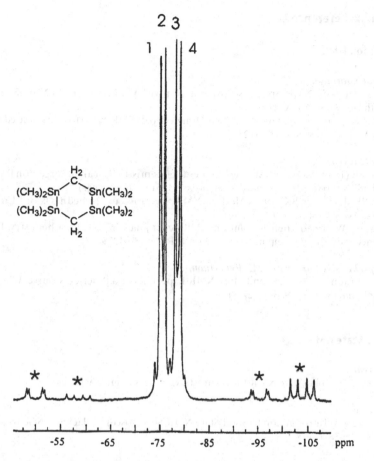

Fig. 54 Central part of the 74.63 MHz spectrum of octamethyl-1,2-3,5-tetrastannacyclohexane. The isotropic shift is –78 ppm, very close to the solution value of –78.5 ppm. Sets of lines marked with an *asterisk* are satellites due to one-bond tin–tin coupling. The linewidths of lines 1 to 4 are 10 Hz

Appendix: Reference List

A Solution NMR

Basic but Thorough
Gunther H (1995) NMR Spectroscopy, 2nd edn. Wiley, Chichester. ISBN 0-471-95199-4 (cloth)/0-471-95261-X (paper)
Friebolin H (2004) Basic One- and Two-Dimensional NMR Spectroscopy, 4th edn. VCH, New York. ISBN 978-3-527-31233-7

More Advanced
Derome AE (1987) Modern NMR Techniques for Chemistry Research. Pergamon Press, Oxford. ISBN 0-08-032514-9 (hardcover)/0-08-032513-0 (softcover)
Sanders JKM, Hunter BK (1993) Modern NMR Spectroscopy, 2nd edn. Oxford University Press, Oxford. ISBN 0-19-855566-0
Claridge TDW (1999) High-Resolution NMR Techniques in Organic Chemistry. Elsevier Science Publishing Company, Amsterdam. ISBN 0-08-042798-7

A Comprehensive Survey of NMR Experiments
Berger S, Braun S (2004) 200 and More NMR Experiments, a Practical Course. Wiley-VCH, Weinheim. ISBN-13: 978-3-527-31067-8

B Solid State NMR

The Classic
Fyfe CA (1983) Solid State NMR for Chemists. CFC Press, Guelph, Canada

Up to Date
Duer MJ (2004) Introduction to Solid State NMR Spectroscopy. Blackwell Science, Oxford. ISBN-13: 978-1-405-10914-7

C Invaluable for Solving Structural Problems

Pretsch E, Bühlmann P, Affolter C (2003) Structure Determination of Organic Compounds. Tables of Spectral Data, 3rd edn. Springer, Berlin, corr. 2nd printing (softcover, with CD-ROM). ISBN-13: 978-3540678151

Part 2: Worked Example and Problems

Readers can obtain a list of answers to the problems by application (by e-mail) to the authors:

Terence N. Mitchell
terence.mitchell@uni-dortmund.de

Burkhard Costisella
burkhard.costisella@uni-dortmund.de

2.1
Section 1

This section contains 35 problems, ordered according to the complexity of the molecule involved. For each problem the following NMR spectra will generally be reproduced:

Proton spectrum with integration
Carbon-13 spectrum with DEPT or APT for multiplicity information
H,H-COSY
H,C-COSY

The 2D spectra are not included for some of the very simplest molecules. Additional information is provided in all cases in the form of
(a) the molecular formula;
(b) IR frequencies corresponding to functional groups present (e.g. OH, NH, C=O, NO$_2$).

Solving the Structures of Organic Molecules

NMR spectroscopy is only one of a series of tools and methods which can be used in order to determine the structures of unknown organic molecules, but since it is by far the most powerful we have decided to concentrate our attention to it in this book. However, in the real world one may not start by running NMR spectra. Preliminary information can be obtained from the following:

Elemental Analysis

If the sample is pure (this can generally be checked by thin layer chromatography or gas chromatography) then the elemental analysis values for carbon, hydrogen and nitrogen can be used to obtain element ratios, provided that C, H, N and O are the only elements present.

Mass Spectrometry

If you are lucky, the ion with the highest mass to charge value will be the molecular ion. However, this is often not the case, as textbooks on mass spectrometry make clear. If it is possible to carry out high resolution mass spectrometry on the molecules in question, and the molecular ion is indeed observed, the exact mass can be used in combination with tables to obtain the molecular formula directly. Alternatively, you can use the internet (http://www.sisweb.com/cgi-bin/mass10.pl) to calculate and plot mass distributions for any molecular fragment you think may be present.

In this book, in order that you can concentrate your attention on the NMR spectra, we shall provide you with the molecular formula in all cases. This in turn provides you with information which can be extremely useful during the process of solving the structure: if the molecule only contains C, H, N and O then you can use the molecular formula to obtain the number of so-called double bond equivalents, i.e. information on the degree of unsaturation. Though there are various formulas which can be devised to do this, we recommend the calculation using the following formula: for a molecule $C_aH_bO_cN_d$, the number of double bond equivalents DBE is calculated as follows

$$DBE = |(2a + 2) - (b - d)|/2$$

Oxygen and any other divalent elements present are ignored. Any other monovalent elements present, such as halogens, are treated as hydrogens, any other tetravalent elements (e.g. Si) as carbons. If other trivalent elements are contained in the molecules (this is rather unlikely for most organic molecules, but trivalent phosphorus is one example) they are treated as nitrogens.

We need to define what is meant by a double bond equivalent: any element–element double bond ($C = C$, $C = O$, $C = N$) count as 1, while triple bonds count as 2. A saturated ring counts as 1, and any double bond present in the ring also counts as 1: thus a benzene ring corresponds to 4 double bond equivalents.

We highly recommend that when solving the problems you make use of a book which contains tables of NMR chemical shifts and coupling constants as well as infrared frequencies. While various suitable texts are available, our preference is the book by Pretsch et al. (see Appendix for details). Though earlier editions are available, we recommend the new 3rd edition, which includes a CD-ROM with NMR prediction software.

These problems have been chosen so that they can be solved using standardized sets of NMR spectra, together with the additional information listed above.

This does not mean that the other techniques discussed in Part 1 are unimportant, but that they will only need to be made use of when the "standard" spectra do not provide sufficient information.

What do we consider to be the standard set of spectra which you should always try to obtain?

The proton spectrum, with integration. This spectrum tells you:
• how many magnetically non-equivalent types of proton are present in the molecule,
• where they absorb (i.e. which type of proton the signals represent), and
• the *relative* numbers of protons. Since we shall give you the total number of protons present, you will be able to calculate the *absolute* numbers of protons of various types in the molecule.

In addition, of course, the coupling constants are often invaluable in determining structural features. Thus for example the coupling constant across three bonds in an olefinic fragment HC=CH is relatively small (8–10 Hz) if the protons are *cis* and much larger (12–16 Hz) if they are *trans*. The magnitude of the corresponding coupling constant in an *aliphatic* fragment HC–CH depends on the **dihedral angle** subtended by the two C–H bonds, this dependence being described semi-quantitatively by the so-called **Karplus equation** (in freely rotating systems such as alkyl chains an average value close to 7 Hz is observed).

A third useful example is provided by aromatic residues, where the coupling constant between *ortho* protons is large (xx–xx Hz), while that between *meta* protons is much smaller (1–3 Hz). The coupling between *para* protons is often of the same magnitude as the linewidth.

The carbon spectrum, both in the broad-band decoupled form and as an APT spectrum.
APT, you will perhaps remember, stands for Attached Proton Test, meaning that this spectrum tells you the multiplicity of the signals (Me, CH₂, CH or quaternary C). These two spectra tell you how many magnetically non-equivalent types of carbon are present in the molecule, but (for the reasons we discussed earlier) we do not use integration to try to find out relative numbers. We shall present APT spectra as follows: CH, CH₃ in negative phase (down), CH₂ and quaternary C in positive phase (up).

You may be told that your NMR laboratory does not routinely use APT spectra but provides DEPT spectra (**Distortionless Enhancement by Polarization Transfer**) instead. This is no problem, as DEPT spectra also provide you with the information you need: just go back and read what we have said about the relative merits of APT and DEPT.

The proton proton COSY spectrum (COSY meaning Correlated SpectroscopY), which tells you directly which protons couple with which. In many cases this information is already available from the proton spectrum, but since multiplets in proton spectra can be quite complicated, even at 400 MHz, the COSY spectra should be recorded as they are very simple to interpret.

The proton carbon correlation spectrum, which tells you directly which signals in the proton spectrum correspond with which signals in the broad-band decoupled carbon spectrum. This information, together with the integration values and the multiplicities obtained from APT (or DEPT), is invaluable in putting together the molecular fragments.

These four NMR spectra will form the basis which you can use to solve the structures (in some cases not all are presented, depending on what information they give). We have naturally arranged the problems on the basis of their molecular complexity, but even very small molecules can have complex proton spectra! All the problems can be solved completely, i.e. including the determination of the isomer involved.

We have used only two different solvents, deuterochloroform (CDCl₃) and hexadeuterodimethylsulfoxide (DMSO-d₆). The former dissolves a large majority of organic molecules, but DMSO must be used for more polar substances. The disadvantage of DMSO is that it is very hygroscopic, so that even if you try hard to keep it dry you may find small signals due to water in your spectra. Look out for these in the spectra in this book: they normally lie at about 3.3 ppm.

Solvent shifts are as follows:

Deuterochloroform: ^1H, residual $CHCl_3$ in $CDCl_3$ 7.26 ppm (singlet), ^{13}C 77 ppm (1:1:1 triplet).

Hexadeuterodimethylsulphoxide: ^1H, residual incompletely deuterated DMSO 2.5 ppm (multiplet), ^{13}C 39.5 ppm (multiplet).

Before we leave you to start on the problems, we feel that it is vital to present a worked example. Though of course you can work as you like, we highly recommend that you try to follow a relatively standard procedure, which will allow you to put together your facts and deductions in a systematic way and thus make it easier for you to arrive at the correct solution in a minimum time.

2.2
Worked Example

First use the molecular formula and the equation given above to calculate the number of double bond equivalents. In this case (remembering to treat bromine as equivalent to hydrogen) the value is 1. The infrared spectrum shows a band at 1641 cm⁻¹, which probably represents the C=C bond stretch, but in this case there can only be a C=C bond present!

This bond is clearly visible in both the proton and carbon spectra. We recommend making a table of the information these give; the tables can be added to as the structure elucidation continues.

Proton NMR:

5.8 ppm	multiplet	1H	olefinic H
5.1 ppm	multiplet	2H	olefinic H
3.4 ppm	triplet	2H	aliphatic H
2.6 ppm	quartet with fine structure	2H	aliphatic H

Carbon-13 NMR:

135.4 ppm	olefinic CH (signal in negative phase)
117.7 ppm	olefinic CH_2
37.2 ppm	aliphatic CH_2
32.2 ppm	aliphatic CH_2

The fact that we have *three* olefinic hydrogens means that our compound is a primary olefin, the fact that the other two carbons are both methylene carbons means that our substituent, bromine, is terminal. Thus the only possibility we have is that we are dealing with 4-bromo-1-butene (try to find another isomer that fits!). But this simple molecules has a highly complex proton spectrum, which can only be interpreted completely (exact chemical shift, coupling constants) by spectrum simulation.

However, we have already obtained the structure, which is

$$H_2C=CH-CH_2-CH_2Br$$

In this case we do not really need the 2D spectra, but we should take the time to look at and interpret them to see how it is done.

H,H Correlation

First draw the diagonal to make interpretation easier. Then label your hydrogens (from left to right: 2, 1, 4, 3). Now look for and (if you like draw) the squares which demonstrate the couplings present. It is immediately obvious that 1 and 2 couple and that 3 and 4 couple (if you did not have the proton assignment, it would be wise to construct another table to show which couplings are present). Since 1 and 2, and also 3 and 4, are separated by three bonds we obviously expect to see coupling.

But we can find two other squares, involving 2 and 3, and 1 and 3. No further square involving 4 is present. When we think about his for a moment, it is obvious (the expert would say "of course 1, 2 and 3 all couple with one another, because it is an allylic system"). If the additional couplings were not present, the number of lines would be much smaller!

C,H Correlation

This time there is no diagonal. You can label the hydrogens again, but how about the carbons? Is it 2, 1, 3, 4 or 2, 1, 4, 3? The beauty of this correlation is that it gives us the answer straight away. The $BrCH_2$ triplet (H-4) corresponds to the highest-field carbon signal, while the H-3 multiplet corresponds to the carbon signal at 37.2 ppm. Thus the order is 2, 1, 3, 4.

If you try to work in the way we have indicated, you should be able to solve all the problems in this section. But do not try to work through them in order: if you feel lost with one of them, just try the next! Good luck!

Worked example : C_4H_7Br
IR : 1641 cm^{-1}
200 MHz, solvent : $CDCl_3$
^{1}H and APT spectra

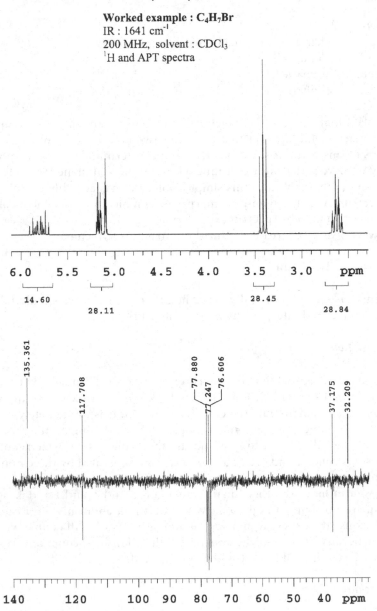

Worked example : C₄H₇Br
200 MHz, solvent : CDCl₃
H,H and C,H correlation

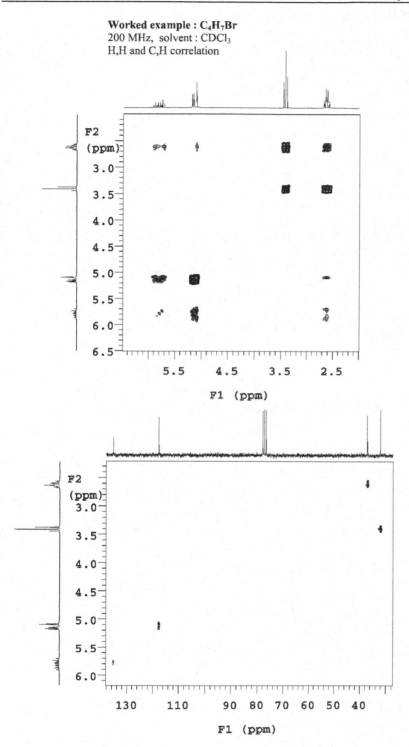

2.3
Problems

For solving the problems, please refer to the worked example (see p. 88) and the detailed data given in the headers of each problem.

Problem 1 : C$_4$H$_6$NCl
IR: 2249 cm^{-1}
200 MHz , solvent : CDCl$_3$
^{1}H and ^{13}C spectra

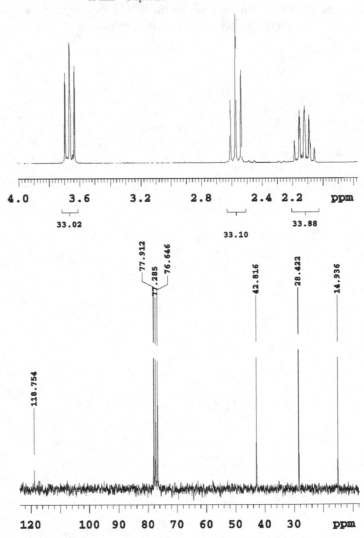

Problem 1 : C_4H_6NCl
200 MHz , solvent : $CDCl_3$
H,H and C,H correlation

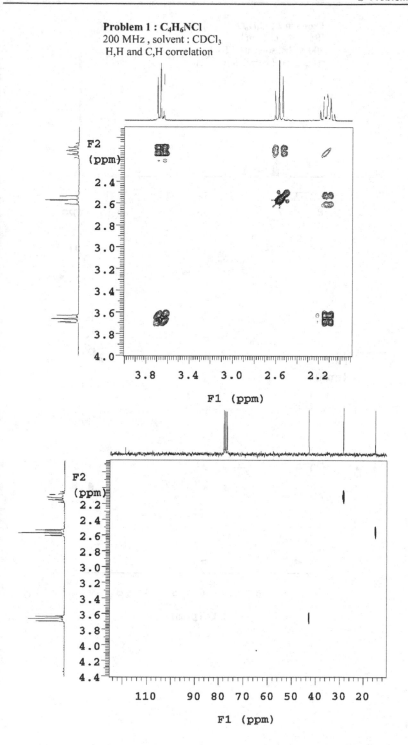

Problem 2 : C₄H₆O

IR: 1654,1691 cm⁻¹

200 MHz , solvent : CDCl₃

¹H and H,H correlation spectra

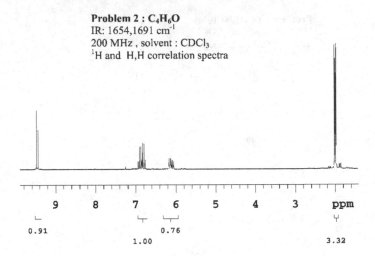

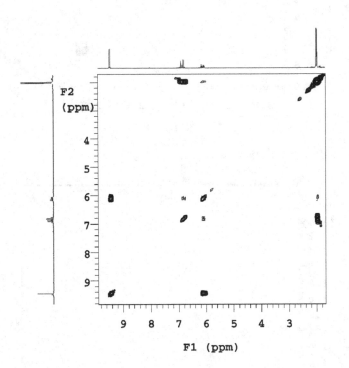

F2
(ppm)

F1 (ppm)

Problem 2 : C_4H_6O
200 MHz , solvent : $CDCl_3$
^{13}C and DEPT spectra

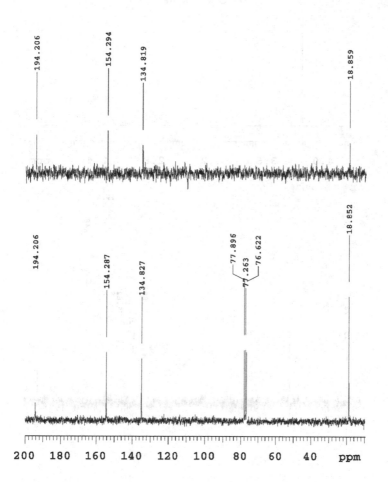

Problem 3 : $C_4H_6O_3$
IR. 1730 cm^{-1} (broad)
200 MHz , solvent : CDCl$_3$
^{1}H and APT spectra

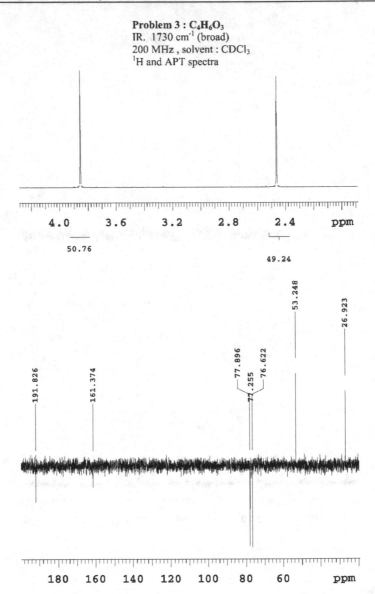

50.76

49.24

Problem 3 : C₄H₆O₃
200 MHz , solvent : CDCl₃
¹³C spectrum and C,H correlation

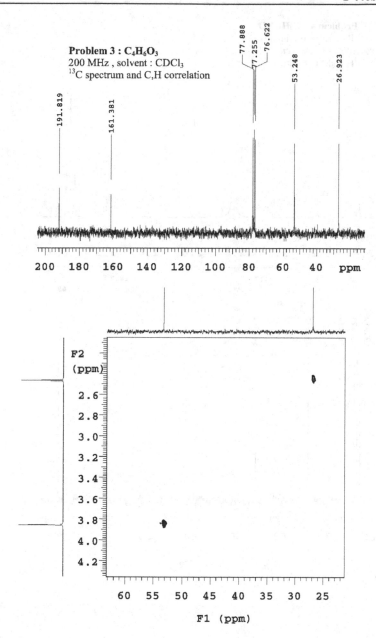

Problem 4 : C₄H₁₁NO

IR: 3350 cm⁻¹ (broad)
200 MHz , solvent : CDCl₃
¹H and ¹³C spectra

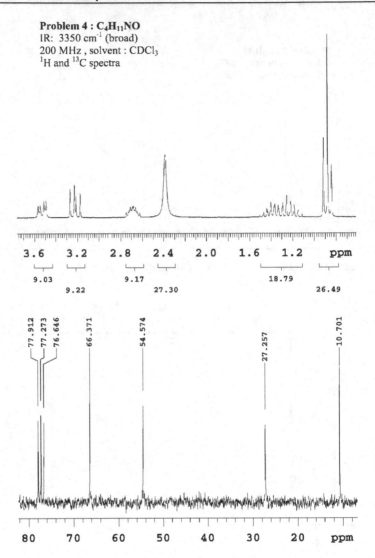

Problem 4 : C$_4$H$_{11}$NO
200 MHz , solvent : CDCl$_3$
H,H and C,H correlation

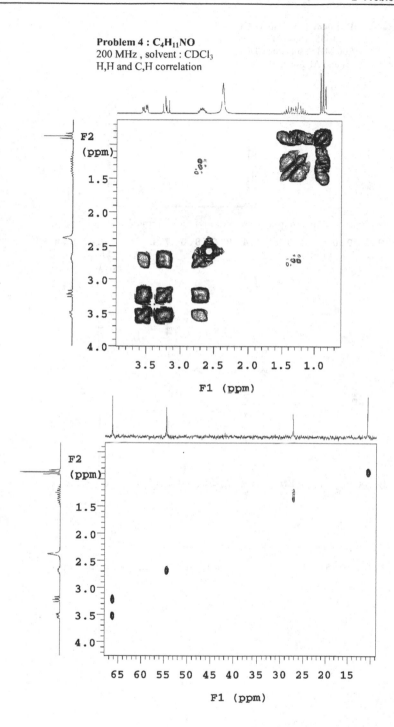

Problem 5 : C$_5$H$_3$N$_2$O$_2$Cl
IR: 1355, 1562 cm^{-1}
200 MHz , solvent : CDCl$_3$
^{1}H and APT spectra

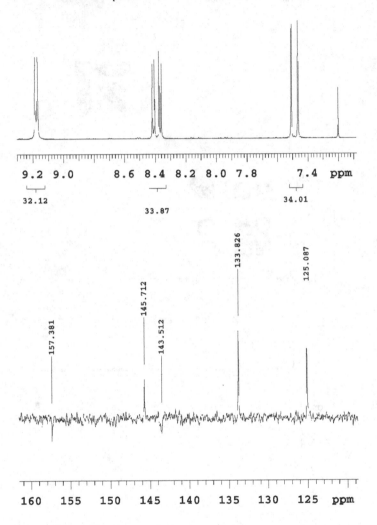

Problem 5 : $C_5H_3N_2O_2Cl$
200 MHz , solvent : $CDCl_3$
H,H and C,H correlation

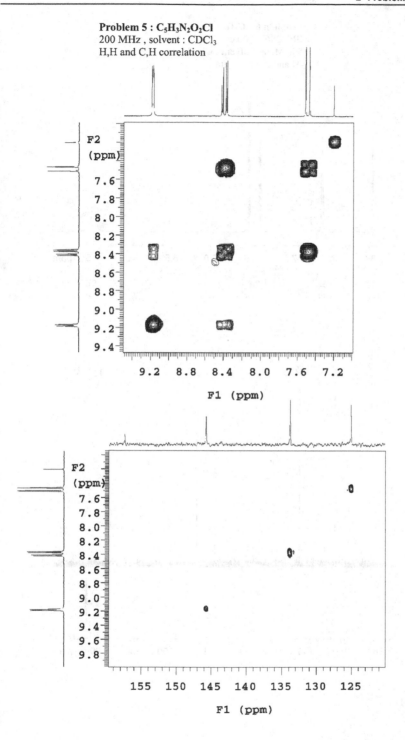

Problem 6 : C₅H₆N₂
IR: 3296, 3360 cm⁻¹
500 MHz , solvent : CDCl₃
¹H and APT spectra

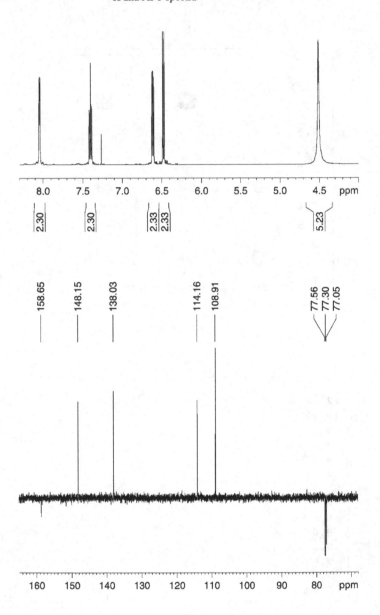

Problem 6 : C₅H₆N₂
500 MHz , solvent : CDCl₃
H, H and C,H correlation

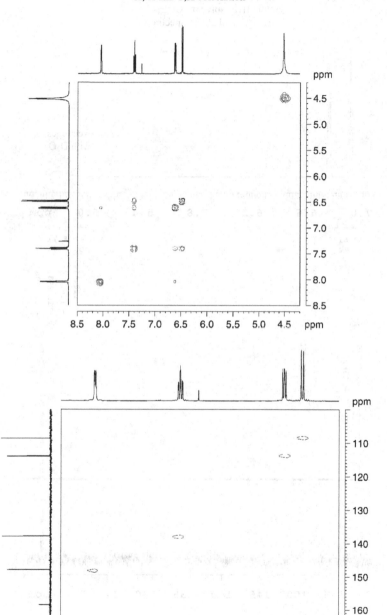

Problem 7 : C$_6$H$_5$OF
^{19}F NMR : δ = -125 ppm
IR: 3215 cm^{-1}
600 MHz , solvent : CDCl$_3$
^{1}H , ^{13}C and DEPT spectra

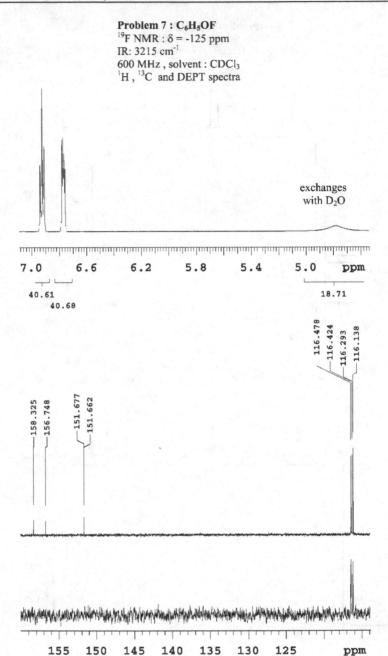

Problem 7 : C₆H₅OF
600 MHz , solvent : CDCl₃
¹³C spectra (expansion)
scale in ppm, peak frequency in Hz

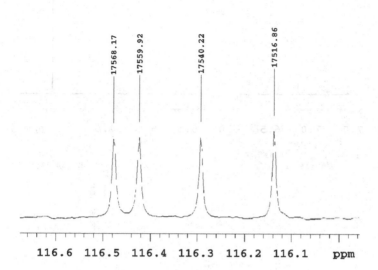

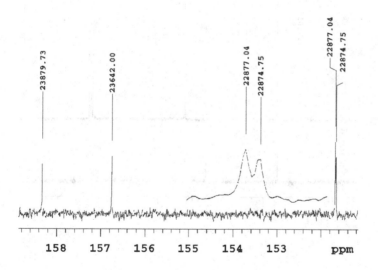

Problem 8 : C₆H₆NOCl

$Problem\ 8 : C_6H_6NOCl$

IR: no bands characteristic of functional groups
200 MHz , solvent : CDCl₃
^1H , ^{13}C and DEPT spectra

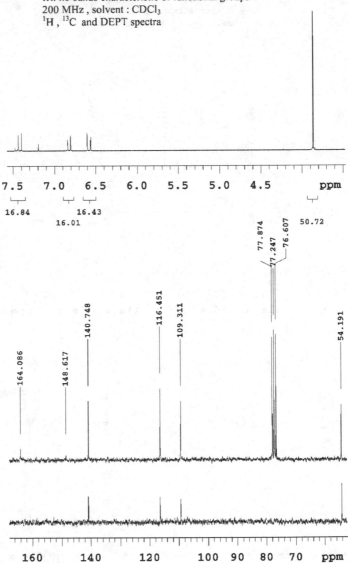

Problem 8 : C₆H₆NOCl
200 MHz , solvent : CDCl₃
H,H and C,H correlation

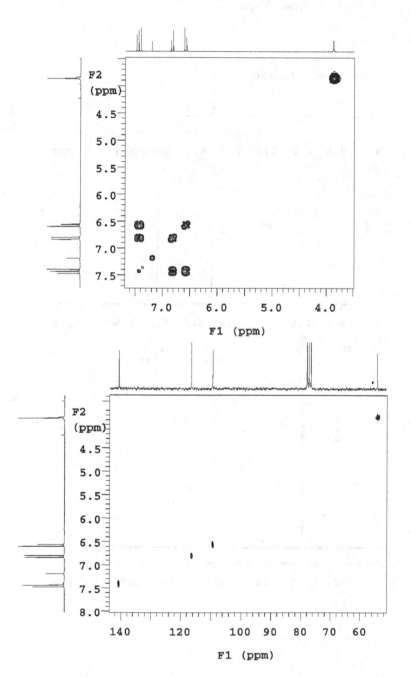

Problem 9 : C₆H₇NO
IR: 3166, 3304, 3360 cm⁻¹
600 MHz , solvent : DMSO-d6
¹H and ¹³C spectra

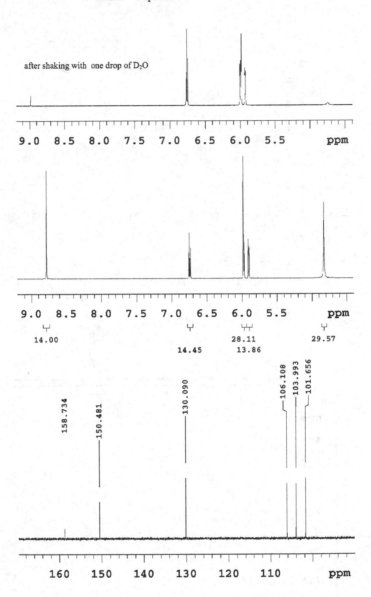

after shaking with one drop of D₂O

Problem 9 : C₆H₇NO
600 MHz , solvent : DMSO-d6
H,H and C,H correlation

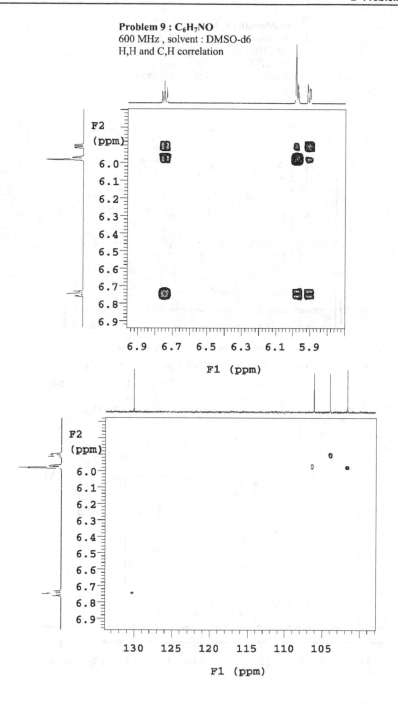

Problem 10 : $C_7H_4N_2O_6$
IR:, 1348, 1545, 1703, 3093 cm^{-1}
200 MHz , solvent : DMSO-d6
^{1}H and ^{13}C spectra
^{1}H expansion : peak frequency in Hz

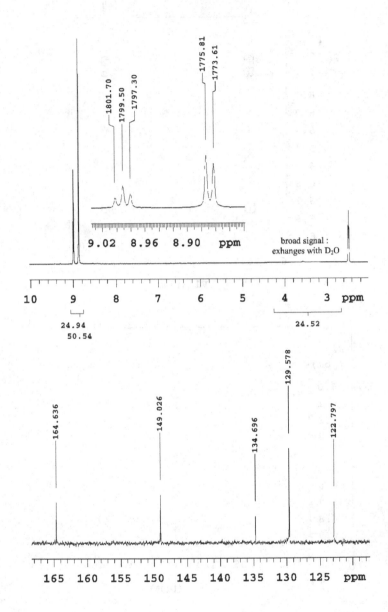

Problem 10 : C$_7$H$_4$N$_2$O$_6$
200 MHz , solvent : DMSO-d6
APT and DEPT spectra

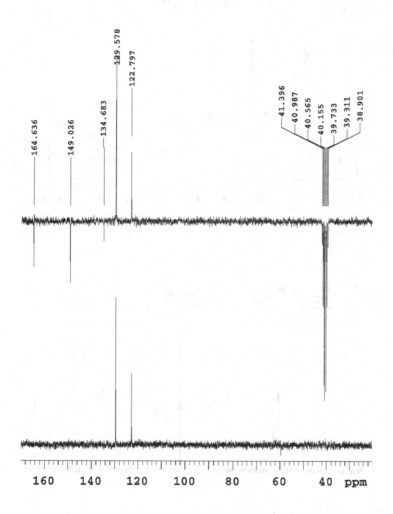

Problem 11 : C₇H₈NCl

IR : 3355, 3430 cm⁻¹
500 MHz , solvent : CDCl₃
¹H , ¹³C and DEPT spectra

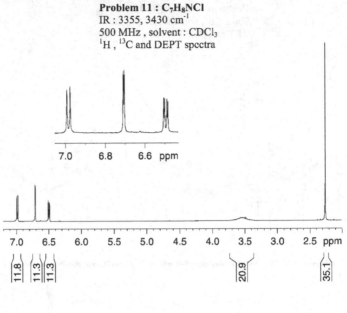

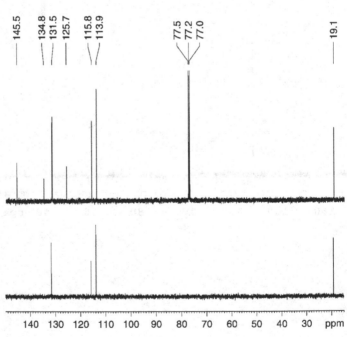

Problem 11 : C₇H₈NCl
solvent : CDCl₃
H,H correlation , 200 MHz
C,H correlation , 500 MHz

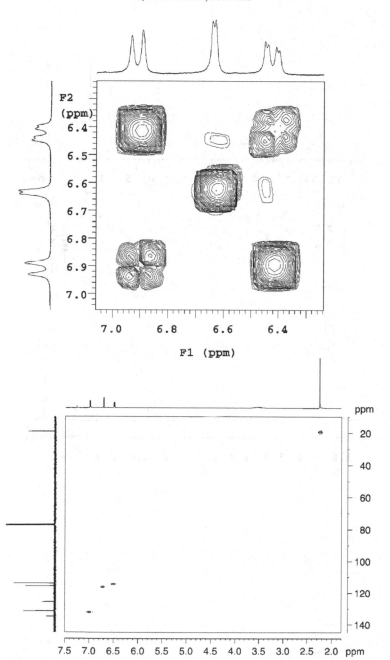

Problem 12 : C$_7$H$_{10}$O$_5$
IR: 1720 (strong), 2930-2970 (broad) cm^{-1}
200 MHz , solvent : DMSO-d6 (with small amount of water : δ = 3.4 ppm broad)
^{1}H spectra with expansion

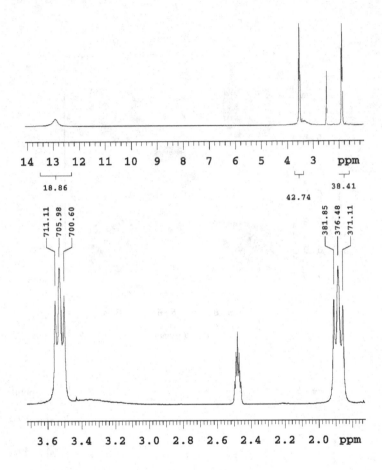

Problem 12 : C₇H₁₀O₅
200 MHz , solvent : DMSO-d6
¹³C and DEPT spectra

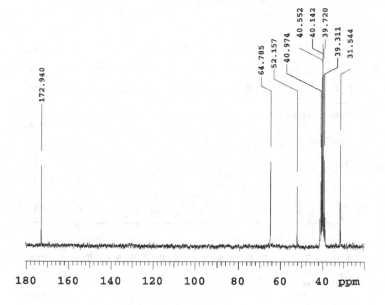

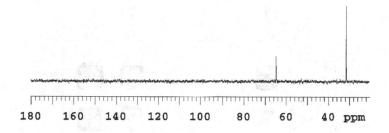

Problem 13 : C_8H_7OBr
IR:1641 cm^{-1}
200 and 600 MHz , solvent : CDCl$_3$
^{1}H spectra and H,H correlation

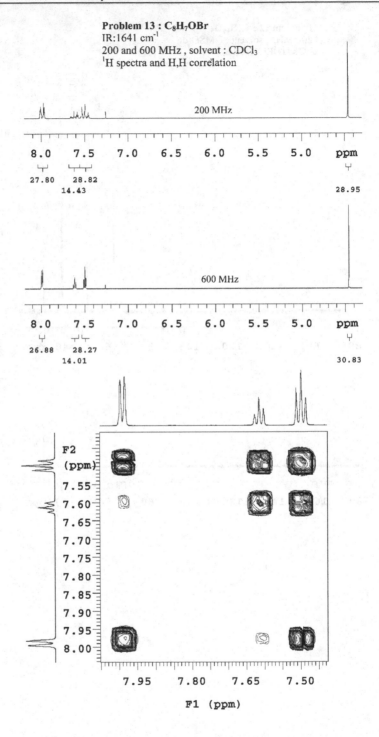

Problem 13 : C$_8$H$_7$OBr
200 MHz , solvent : CDCl$_3$
^{13}C spectra and C,H correlation

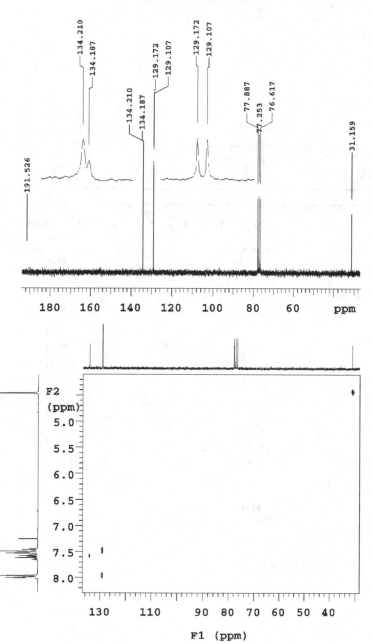

Problem 14 : $C_8H_{10}N_2O$
IR: 1681, 3093, 3159, 3279 cm^{-1}
200 MHz , solvent : CDCl3 (top) and DMSO-d6 (below)
^{1}H spectra

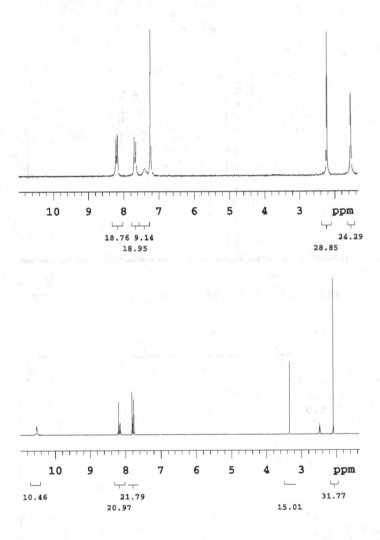

Problem 14 : C₈H₁₀N₂O
200 MHz , solvent : DMSO-d6
¹³C and DEPTspectra

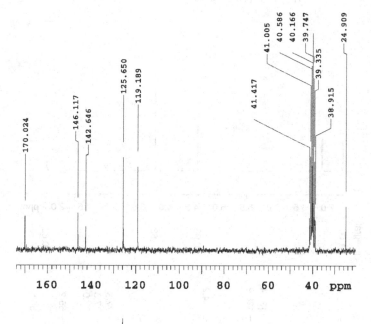

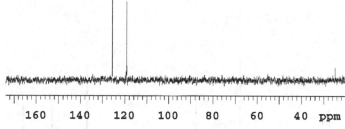

Problem 15 : C₈H₁₀O₂
IR: 3365 (very broad) cm⁻¹
500 MHz , solvent : CDCl₃
¹H , ¹³C and DEPT spectra

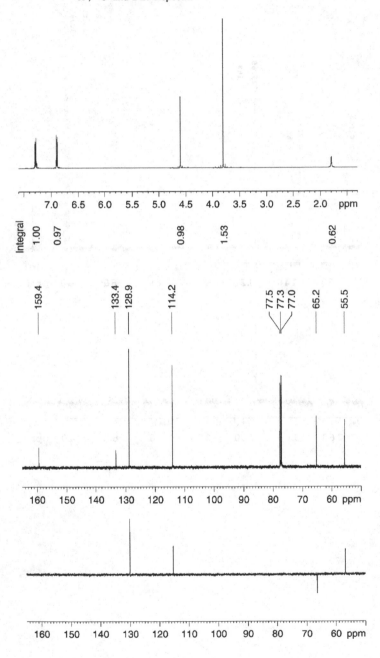

Problem 15 : C$_8$H$_{10}$O$_2$
500 MHz , solvent : CDCl$_3$
H,H and C,H correlation

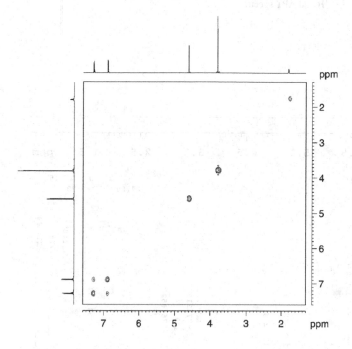

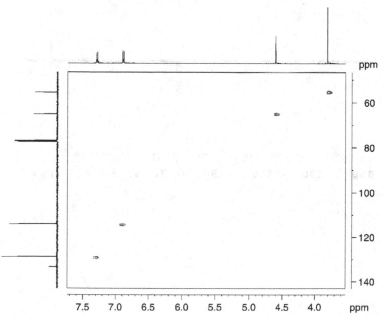

Problem 16 : C₈H₁₁N
IR . 1637, 2214 cm⁻¹
200 MHz , solvent : CDCl₃
¹H and APT spectra

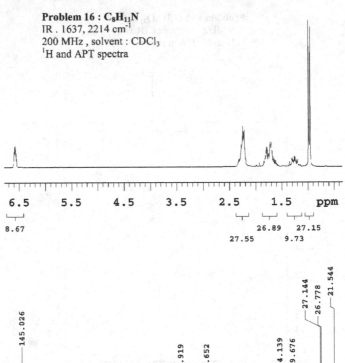

6.5   5.5   4.5   3.5   2.5   1.5   ppm

8.67   26.89   27.15
                                          27.55   9.73

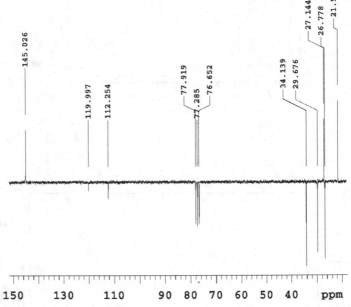

150   130   110   90 80 70 60 50 40   ppm

Problem 16 : C₈H₁₁N
200 MHz , solvent : CDCl₃
H,H and C,H correlation

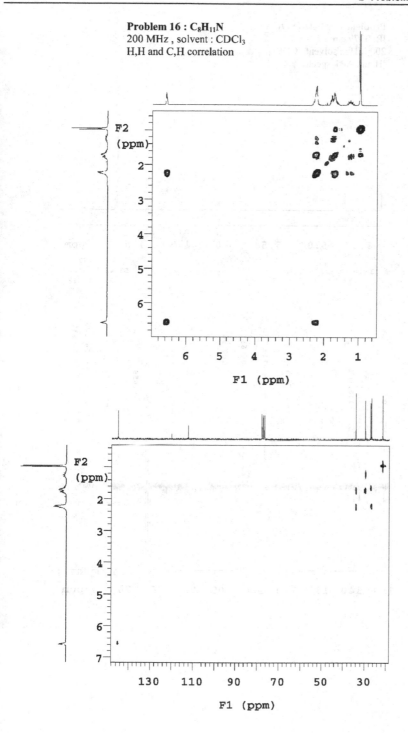

Problem 17 : C₈H₁₁NO₂
IR: 3379 cm⁻¹
200 MHz , solvent : CDCl₃
¹H and APT spectra

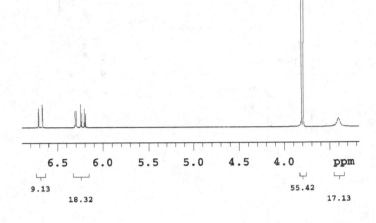

9.13
18.32
55.42
17.13

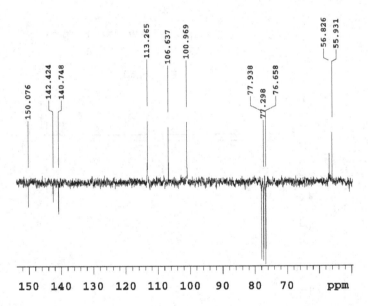

Problem 17 : $C_8H_{11}NO_2$
200 MHz , solvent : $CDCl_3$
H,H and C,H correlation

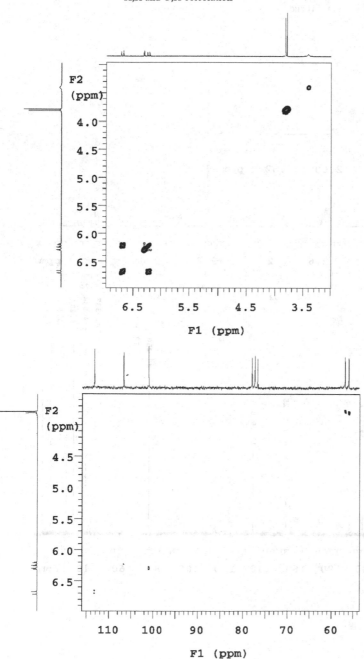

F2
(ppm)

4.0

4.5

5.0

5.5

6.0

6.5

6.5   5.5   4.5   3.5

F1 (ppm)

F2
(ppm)

4.5

5.0

5.5

6.0

6.5

110   100   90   80   70   60

F1 (ppm)

Problem 18 : C₈H₁₂O
IR: 1680 cm⁻¹
600 MHz , solvent : CDCl₃
¹H and ¹³C spectra

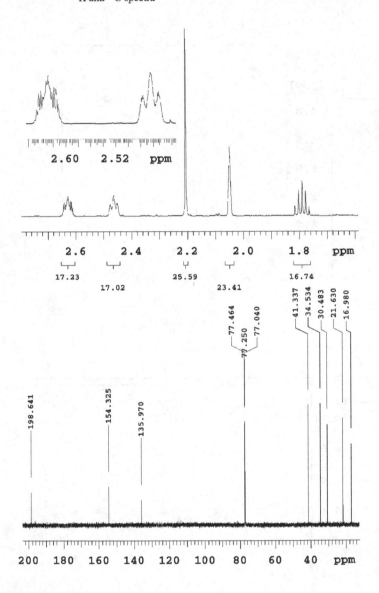

Problem 18 : C₈H₁₂O
600 MHz , solvent : CDCl₃
H,H and C,H correlation

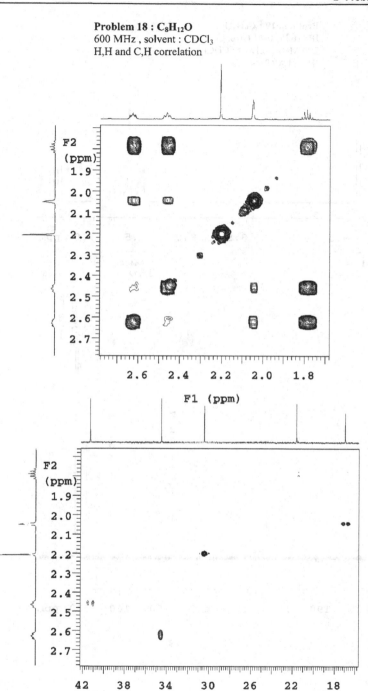

Problem 19 : C₉H₈O
IR: 1625, 1681 (strong) cm⁻¹
200 MHz , solvent : CDCl₃
¹H and APT spectra

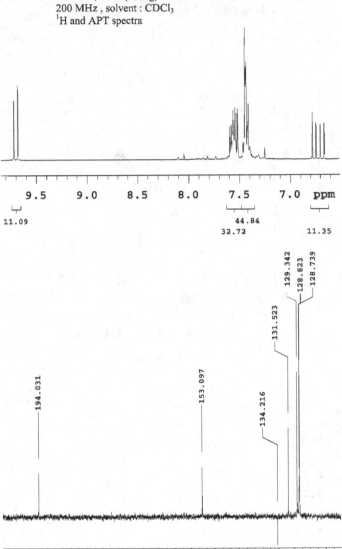

Problem 19 : C₉H₈O
200 MHz , solvent : CDCl₃
H,H correlation anda part of C,H correlation

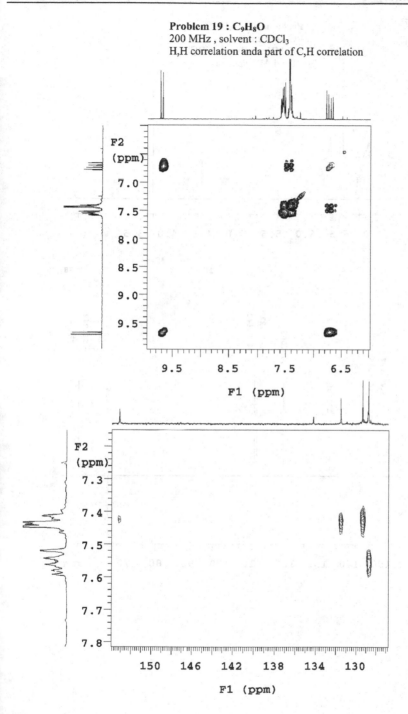

Problem 20 : C$_9$H$_{12}$O$_3$
IR: 3480 cm^{-1}
400 MHz , solvent : CDCl$_3$
^{1}H and APT spectra

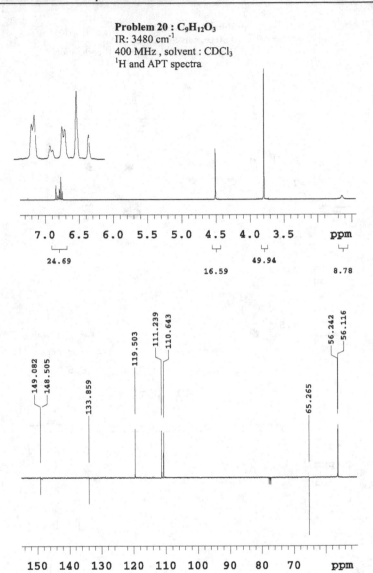

Problem 20 : C9H12O3
400 MHz , solvent : CDCl3
13C spectra and C,H correlation

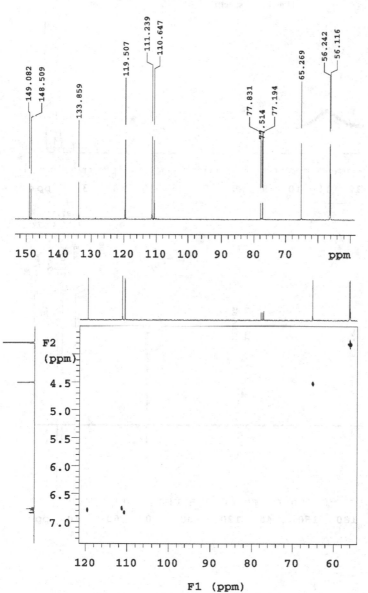

Problem 21 : C₉H₁₄O₂
IR: 1697, 2900 (broad) cm⁻¹
400 MHz , solvent : CDCl₃
¹H and APT spectra

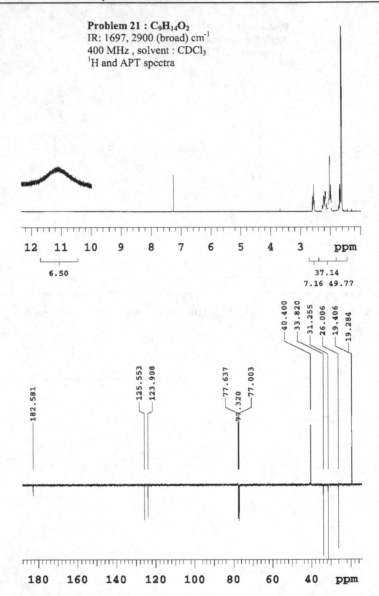

Problem 21 : C$_9$H$_{14}$O$_2$
400 MHz , solvent : CDCl$_3$
H,H and C,H correlation

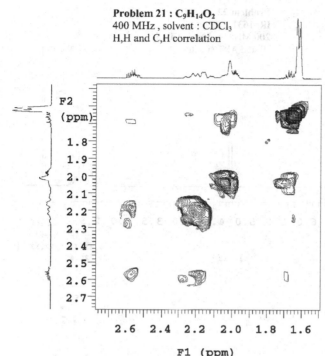

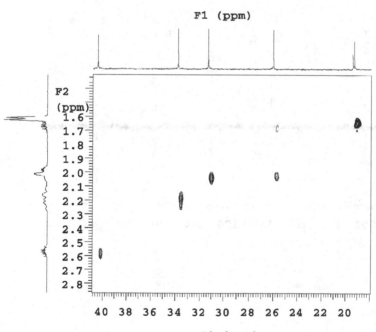

Problem 22 : C₁₀H₁₄O
IR: 1631, 1732 cm⁻¹
200 MHz , solvent : CDCl₃
¹H and APT spectra

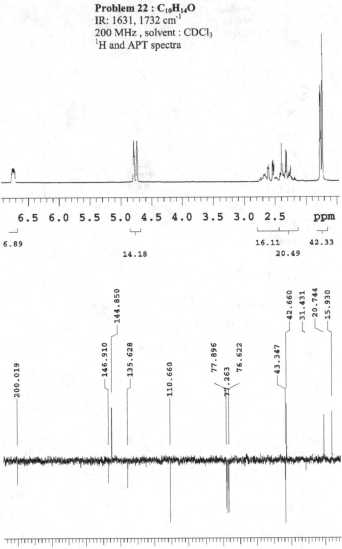

Problem 22 : C$_{10}$H$_{14}$O
200 MHz , solvent : CDCl$_3$
H,H and C,H correlation

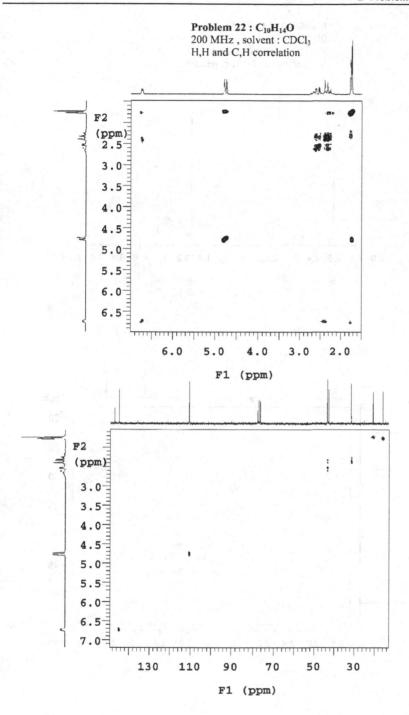

Problem 23 : $C_{10}H_{16}O_2$
IR: 1725 cm^{-1}
500 MHz , solvent : CDCl$_3$
^{1}H spectrum and C,H correlation

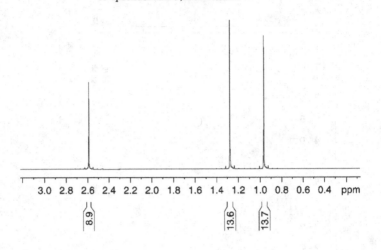

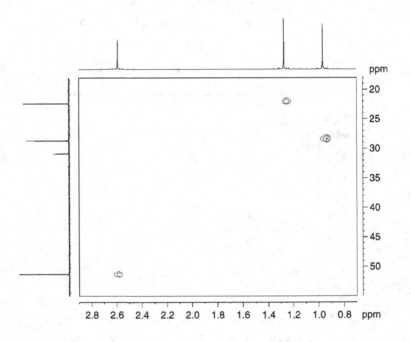

Problem 23 : C₁₀H₁₆O₂
500 MHz , solvent : CDCl₃
¹³C and APT spectra

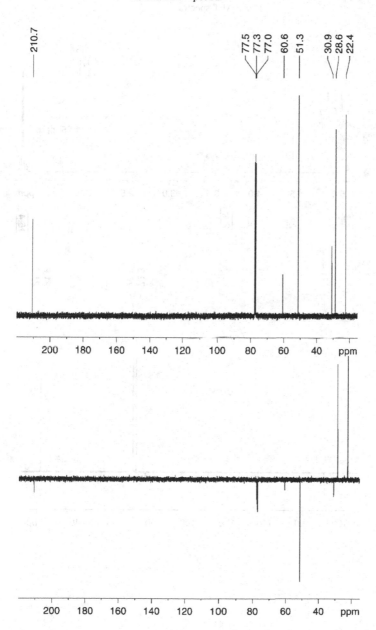

Problem 24 : C₁₀H₁₈O₆
IR: 1730(strong), 3340(broad) cm⁻¹
500 MHz , solvent : CDCl₃
¹H and APT spectra

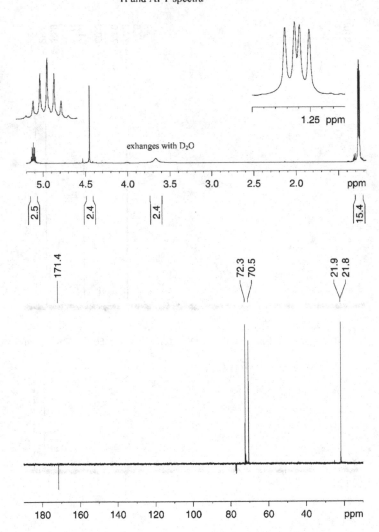

Problem 24 : C₁₀H₁₈O₆
500 MHz , solvent : CDCl₃
H,H and C,H correlation

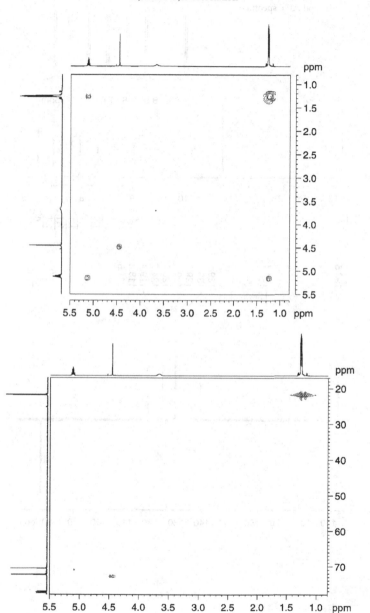

Problem 25 : $C_{11}H_8O_2$
IR: 1670(strong), 3360 cm⁻¹
500 MHz , solvent : CDCl₃
¹H and APT spectra

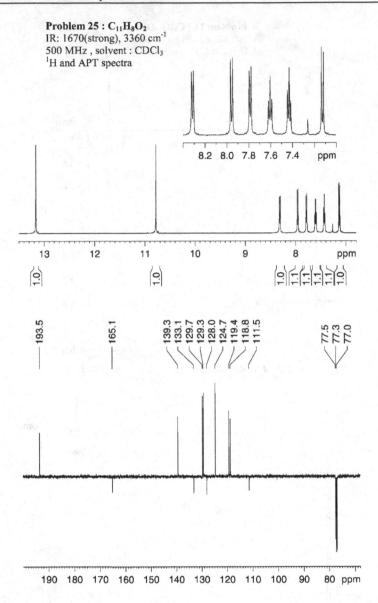

Problem 25 : $C_{11}H_8O_2$
500 MHz , solvent : $CDCl_3$
H,H and C,H correlation

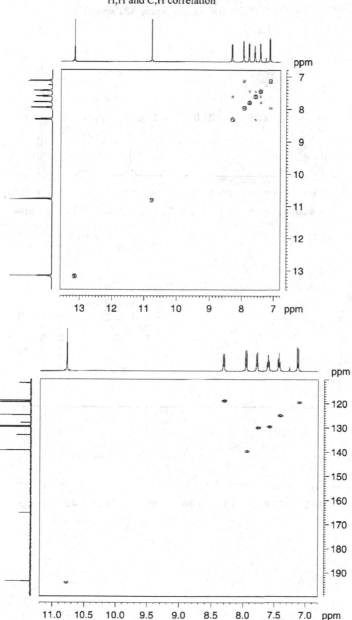

Problem 26 : C₁₁H₁₈O₄
IR: 1732 cm⁻¹
200 MHz , solvent : CDCl₃
¹H with expansion and APT spectra

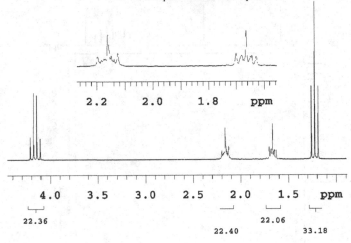

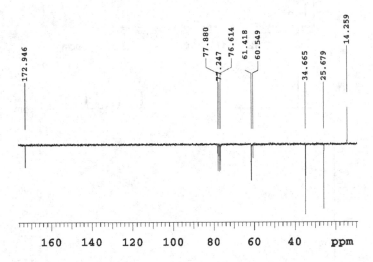

Problem 26 : C₁₁H₁₈O₄
200 MHz , solvent : CDCl₃
H,H and C,H correlation

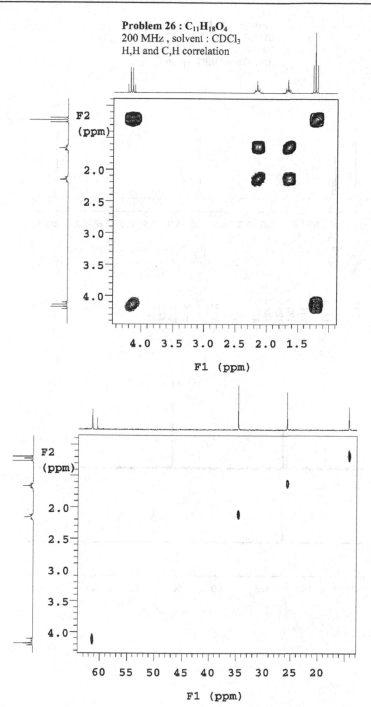

Problem 27 : C₁₂H₁₁NO₃
IR: 1631, 1714, 2216, 3230 cm⁻¹
500 MHz , solvent : CDCl₃
¹H , ¹³C and DEPT spectra

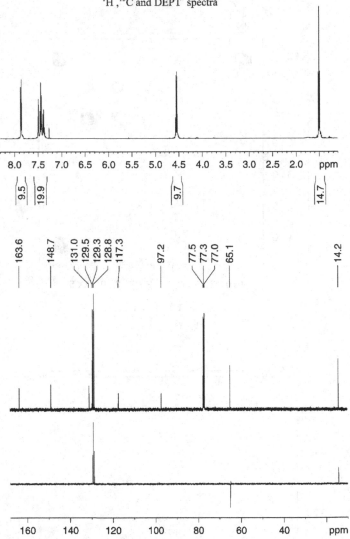

147

Problem 27 : C₁₂H₁₁NO₃
500 MHz , solvent : CDCl₃
H,H and C,H correlation

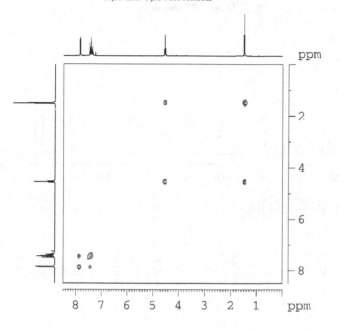

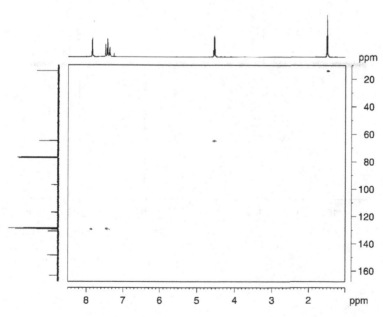

Problem 28 : $C_{12}H_{20}O_2$
IR: 1680, 1740 cm^{-1}
500 MHz , solvent : CDCl$_3$
^{1}H and APT spectra

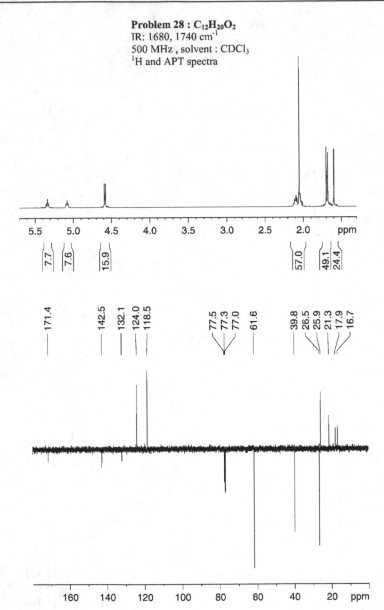

Problem 28 : C$_{12}$H$_{20}$O$_2$
500 MHz , solvent : CDCl$_3$
H,H and C,H correlation

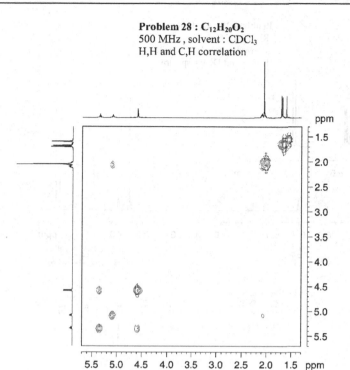

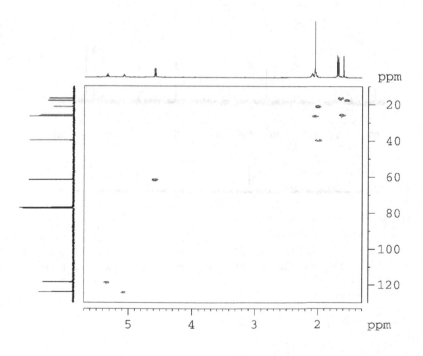

Problem 29 : C₁₃H₂₀O

$\text{Problem 29 : } C_{13}H_{20}O$

IR: 1620, 1674 cm⁻¹

400 MHz , solvent : CDCl₃

¹H ,¹³C and DEPT spectra

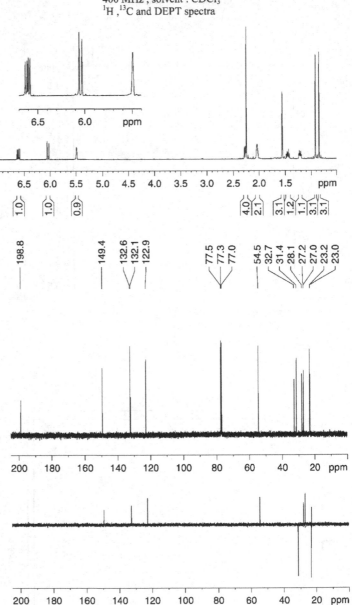

Problem 29 : C₁₃H₂₀O

Problem 29 : $C_{13}H_{20}O$
400 MHz , solvent : CDCl₃
H,H and C,H correlation

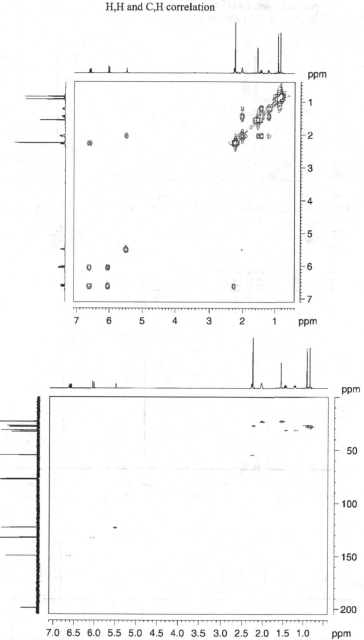

Problem 30 : C₁₄H₁₈N₂

IR: no bands characteristic of functional groups
500 MHz , solvent : CDCl₃
¹H , ¹³C and APT spectra

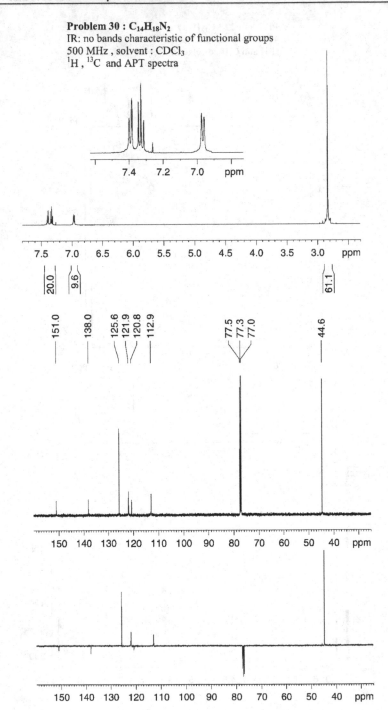

Problem 30 : C₁₄H₁₈N₂
500 MHz , solvent : CDCl₃
H,H and C,H correlation

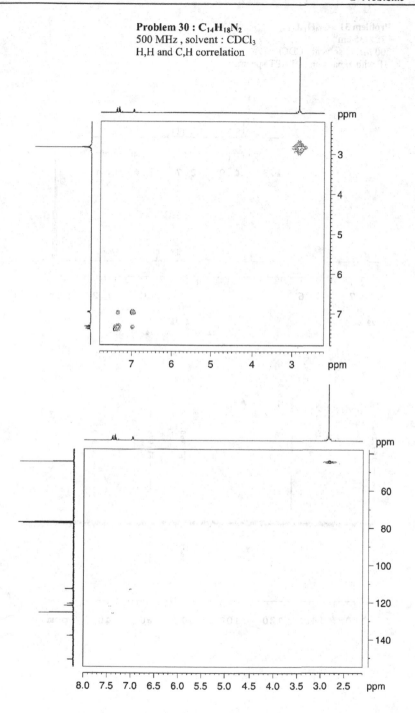

Problem 31 : C$_{14}$H$_{18}$O$_4$
IR: 1735 cm^{-1}
200 MHz , solvent : CDCl$_3$
^{1}H with expansion and APT spectra

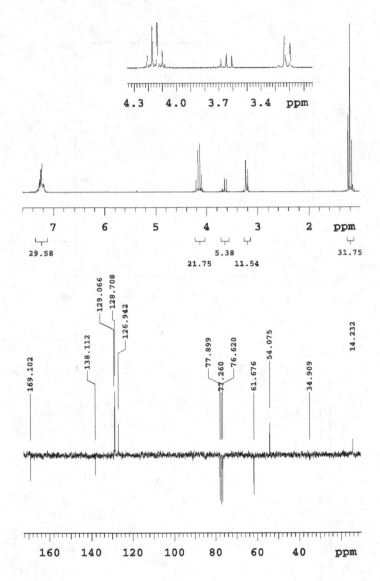

Problem 31 : C₁₄H₁₈O₄
200 MHz , solvent : CDCl₃
H,H and C,H correlation

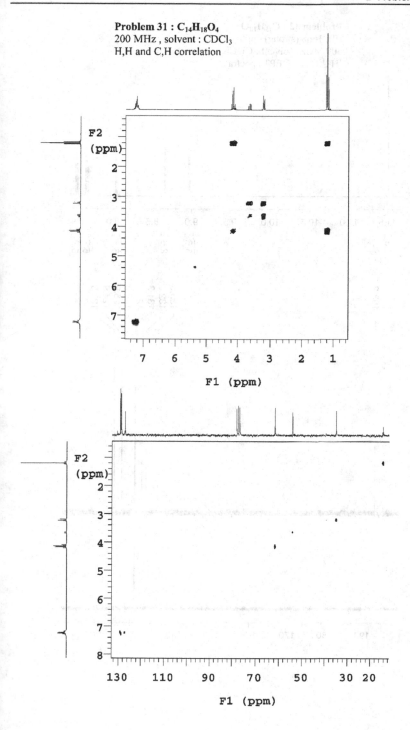

Problem 32 : C₁₅H₁₀O

IR:, 1668 (strong) cm⁻¹
500 MHz , solvent : CDCl₃
¹H ,¹³C and DEPT spectra

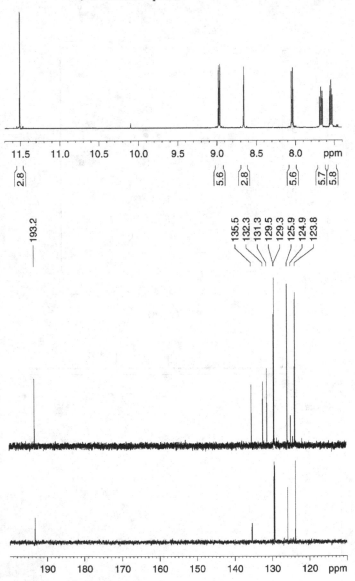

Problem 32 : C$_{15}$H$_{10}$O
500 MHz , solvent : CDCl$_3$
H,H correlation
aromatic part of the C,H correlation

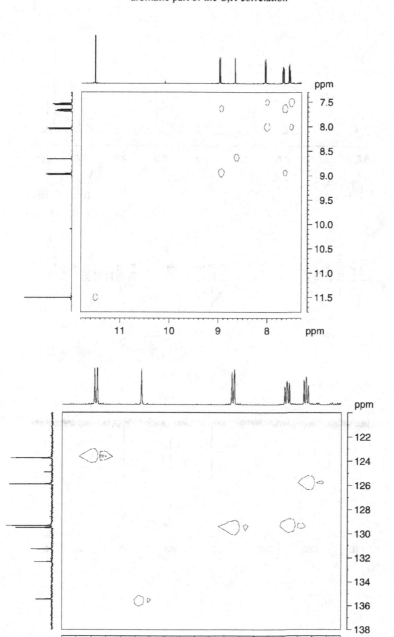

Problem 33 : C$_{15}$H$_{26}$O
IR: 1668, 3366 (broad) cm^{-1}
500 MHz , solvent : CDCl$_3$
^{1}H and APT spectra

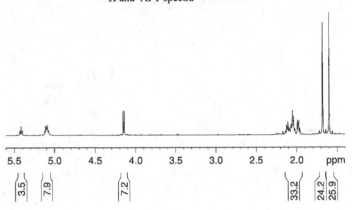

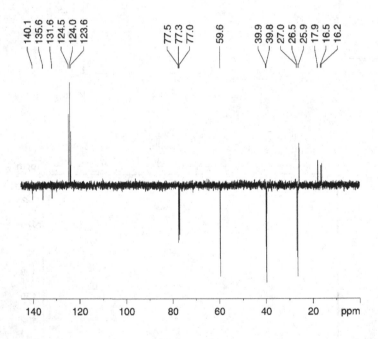

Problem 33 : C₁₅H₂₆O
500 MHz , solvent : CDCl₃
H,H and C,H correlation

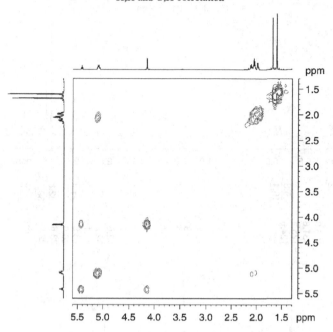

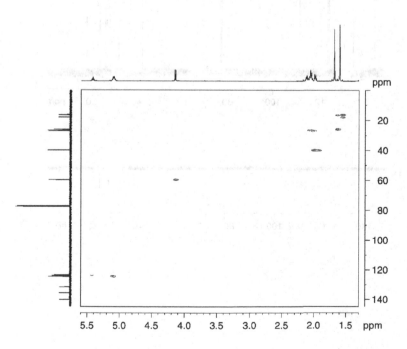

Problem 34 : C₁₅H₂₆O
IR: 1675,3406 (broad) cm⁻¹
500 MHz , solvent : CDCl₃
¹H ,¹³C and DEPT spectra

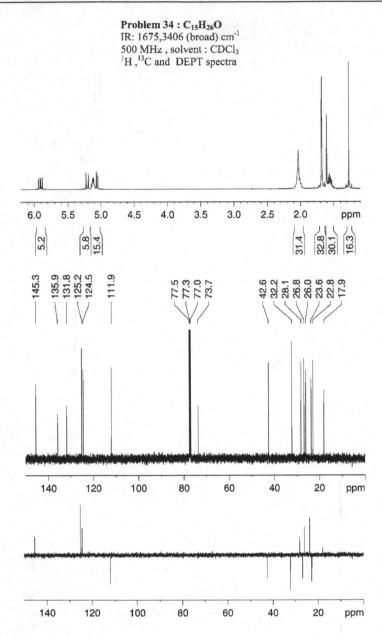

Problem 34 : C₁₅H₂₆O
500 MHz , solvent : CDCl₃
H,H and C,H correlation

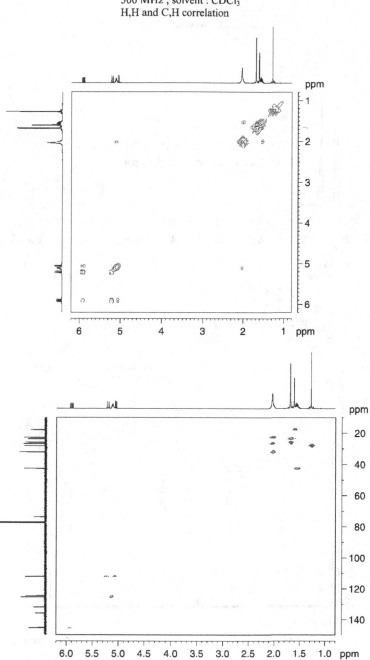

Problem 35 : C₁₈H₁₉OP

^{31}P NMR : δ = 8.6 ppm
IR: 1195, 2194 cm^{-1}
200 MHz , solvent : CDCl₃
^{1}H and ^{13}C spectra

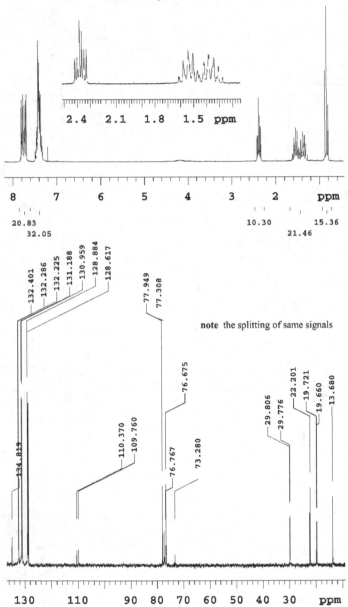

note the splitting of same signals

Problem 35 : C$_{18}$H$_{19}$OP
200 MHz , solvent : CDCl$_3$
APT and DEPT spectra
expansion : peak frequency in Hz

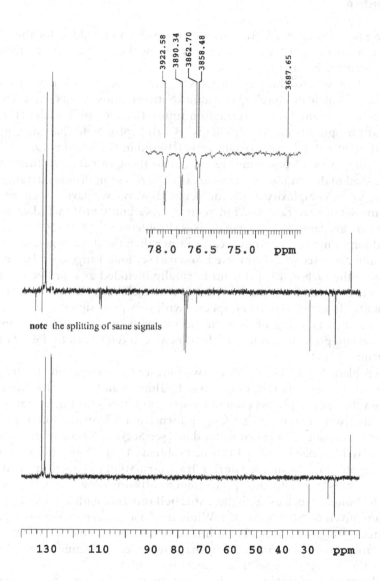

note the splitting of same signals

2.4
Section 2

Introduction

The 15 problems in this section (numbers 36–50) were added for the second edition of the book. They do not exactly follow the scheme set out above for the problems in Section 1.

First of all we have three problems where the structure is known. Here you are asked to calculate coupling constants between phosphorus and carbon or hydrogen (Problem 36) and relaxation times T_1 for carbon nuclei (Problem 37) and phosphorus nuclei (Problems 38a and 38b). Note that the equation you will require for T_1 calculations can be found in Fig. 10 on p. 19

The remaining 12 problems are similar to those in Part 1 in that you are again asked to determine the structures of a series of molecules, arranged according to the complexity of the molecule. However, we have not always used the same set of spectra as used in Section 1. Naturally proton and carbon-13 spectra are available. But in several cases (Problems 39–43, 48–50) we have recorded long-range correlation spectra rather than the direct ones. Refer back to Section 2.6 to see what is involved. You will see long-range correlation spectra where the carbon-13 spectrum normally included as a projection in 2D spectra is missing. The reason for this is that such long-range correlation experiments often give 1D carbon spectra with very poor signal-to-noise ratios. But the computer still generates the carbon ppm axis, so in fact the absence of the carbon projection is no problem, as we can refer back to the 1D carbon spectrum.

In Problem 40 we have recorded two long-range correlation spectra using delays set for two different long-range J values, 8 and 2 Hz. The differences between the spectra give us even more structurally-relevant information.

Another way of trying to get more information is by using different methods for processing the experimental data (see Section 1.1.3 and in particular Fig. 6). We have demonstrated this in Problems 43–47. Normally we just ask the computer to do a simple Fourier Transformation (using a typed-in command such as "ft"). But the spectrometer software allows us to use several other techniques, such as applying a sine bell function with a defined parameter (here given as "sb = value x"). While the baseline across the multiplets is no longer straight, additional coupling information becomes clearly visible.

In Problems 41, 43, 45 and 50 we have recorded one-dimensional NOESY and TOCSY spectra (see Section 1.1.6.2 for details).

Problem 41 confronts you with a phenomenon which we have referred to briefly in Section 1.1.6: a negative NOE signal. The point to note is that this is real; do not worry about how it arises!

In Problem 45 we have irradiated the same two signals in the NOESY and TOCSY experiments. In the NOESY spectra the protons which "answer" (show a clearly magnitude-enhanced signal) are those which are closer to the irradiated protons. Of the four multiplets between 7 and 7.5 ppm, only ONE is enhanced for each irradiation.

The TOCSY spectra, as we have said above, contain enhancements connected with spin systems and not with distances. So the "answers" to the irradiation are different. You should be able to interpret them when you have solved the problem!

In Problem 42 we have included the results of several homodecoupling experiments (Section 1.1.1), which simplify multiplets and allow ready determination of coupling constants in otherwise complex multiplets.

In Problem 49 we recorded the carbon-13 spectrum using a relaxation delay of 25 sec; with the shorter delays we tend to use routinely the signals due to the quaternary carbons would have been almost invisible!

In Problem 50 we start by showing you how the proton spectrum varies depending on the spectrometer's magnetic field. The increased **spectral dispersion** at 600 MHz makes quite a difference! The multiplets look completely different, as you can see better in the expansions. Even at 600 MHz spectrum simulation will be required for a complete determination of the coupling constants, but we can simplify the multiplets quite a bit using NOESY and TOCSY.

Problem 36 : $C_{11}H_{11}Cl_4O_5P$
Determination of P-C and C-H coupling constants

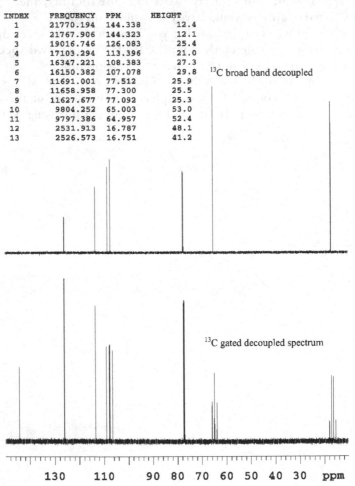

600 MHz , solvent : $CDCl_3$
^{13}C broad band decoupled and ^{13}C gated decoupled spectra
Assign all signals and calculate and tabulate coupling constants $^nJ(P-C)$ and $^nJ(C-H)$

INDEX	FREQUENCY	PPM	HEIGHT
1	21770.194	144.338	12.4
2	21767.906	144.323	12.1
3	19016.746	126.083	25.4
4	17103.294	113.396	21.0
5	16347.221	108.383	27.3
6	16150.382	107.078	29.8
7	11691.001	77.512	25.9
8	11658.958	77.300	25.5
9	11627.677	77.092	25.3
10	9804.252	65.003	53.0
11	9797.386	64.957	52.4
12	2531.913	16.787	48.1
13	2526.573	16.751	41.2

^{13}C broad band decoupled

^{13}C gated decoupled spectrum

130   110   90 80 70 60 50 40 30   ppm

Problem 36 : $C_{11}H_{11}Cl_4O_5P$

^{13}C gated decoupled spectrum. Expanted multiplets : peak frequencies in Hz

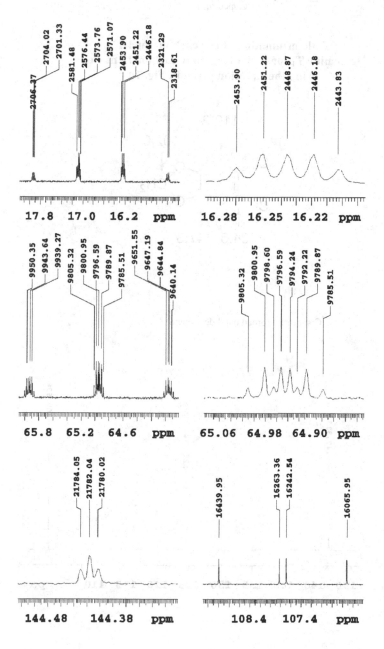

Problem 37 : C₇H₆O₂

200 MHz , solvent : CDCl₃ (degassed)

¹³C spectrum

T₁ determination of the carbon nuclei
Determine T₁ for each of the four carbon-13 nuclei
using the equation given in Fig. 10

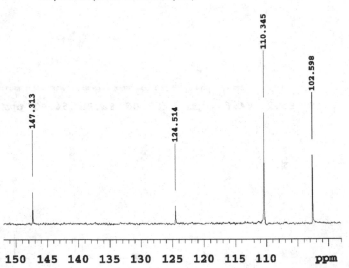

¹³C spectrum (broad band decoupled)

Problem 37 : $C_7H_6O_2$

200 MHz , solvent : $CDCl_3$ (degassed)
T_1 determination of the carbon atoms
by inversion recovery experiment ;
selected ^{13}C spectra

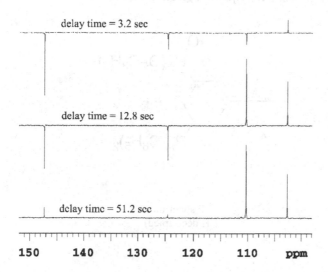

delay time = 3.2 sec

delay time = 12.8 sec

delay time = 51.2 sec

Experimental data for T_1 determination
by exponential data analysis

variable delay time [sec]	δ=147.3 signal intensity	δ=124.5 signal intensity	δ=110.3 signal intensity	δ=102.6 signal intensity
0.1	-10.7	-9.02	-14.1	-5.52
0.2	-10.9	-8.73	-13.7	-5.35
0.4	-10.9	-8.34	-12.5	-4.82
0.8	-11.0	-8.06	-10.5	-3.60
1.6	-10.6	-8.01	-7.17	-1.31
3.2	-10.1	-7.87	-1.82	2.08
6.4	-8.99	-7.06	4.98	5.30
12.8	-6.98	-5.63	10.9	7.15
25.6	-3.44	-2.99	13.2	7.55
51.2	1.69	1.10	12.0	7.23
102.4	7.22	5.89	10.9	6.91
204.8	10.6	9.11	11.3	7.03

Problem 38a : $C_{28}H_{36}O_3P_2$
81.015 MHz , solvent : $CDCl_3$ (not degassed)
^{31}P spectrum

T_1 determination of the phosphorus nuclei
Determine T_1 for the two phosphorus nuclei using the equation given in Fig. 10

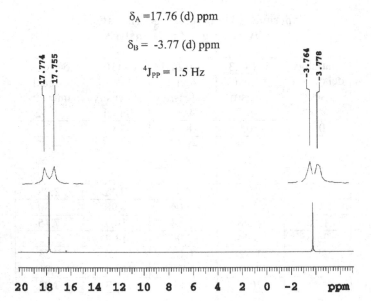

^{31}P spectrum
(^1H decoupled)

$\delta_A = 17.76$ (d) ppm

$\delta_B = -3.77$ (d) ppm

$^4J_{PP} = 1.5$ Hz

Problem 38a : $C_{28}H_{36}O_3P_2$
81.015 MHz , solvent : $CDCl_3$ (not degassed)
T_1 determination of the phosphorus nuclei
by inversion recovery experiment ;
selected ^{31}P spectra

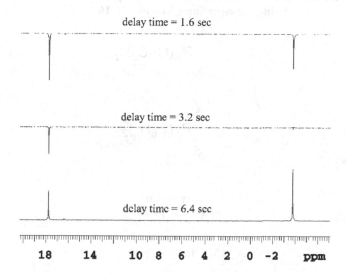

delay time = 1.6 sec

delay time = 3.2 sec

delay time = 6.4 sec

Experimental data for T_1 determination
by exponential data analysis

variable delay time [sec]	$\delta = 17.774$ signal intensity	$\delta = 17.755$ signal intensity	$\delta = -3.764$ signal intensity	$\delta = -3.778$ signal intensity
0.025	-48.3	-50.5	-48.1	-48.1
0.05	-48.6	-49.4	-46.7	-47.5
0.1	-48.3	-48.5	-45.9	-46.4
0.2	-47.2	-47.2	-44.6	-44.2
0.4	-44.8	-44.8	-40.4	-41.1
0.8	-38.9	-38.9	-33.6	-34.2
1.6	-28.3	-28.3	-20.8	-21.0
3.2	-10.6	-10.6	-0.41	-0.33
6.4	13.9	13.9	25.0	24.8
12.8	37.8	37.8	43.6	43.6
25.6	49.8	49.8	48.7	49.2
51.2	51.6	51.6	48.5	49.0

Problem 38b : C$_{28}$H$_{36}$O$_3$P$_2$S
81.015 MHz , solvent : CDCl$_3$ (not degassed)
^{31}P spectrum

T$_1$ determination of the phosphorus nuclei
Determine T$_1$ for the two different phosphorus nuclei
Using the equation given in Fig. 10

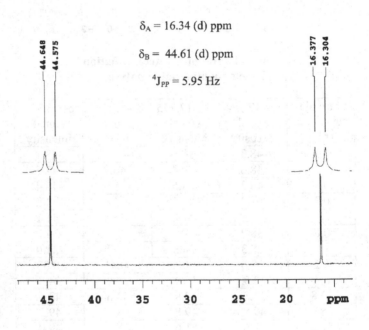

^{31}P spectrum
(^{1}H decoupled)

$\delta_A = 16.34$ (d) ppm

$\delta_B = 44.61$ (d) ppm

$^4J_{PP} = 5.95$ Hz

Problem 38b : C$_{28}$H$_{36}$O$_3$P$_2$S
81.015 MHz , solvent : CDCl$_3$ (degassed)
T$_1$ determination of the phosphorus nuclei
by inversion recovery experiment ;
selected ^{31}P spectra

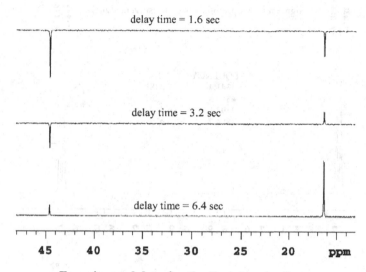

delay time = 1.6 sec

delay time = 3.2 sec

delay time = 6.4 sec

45   40   35   30   25   20   ppm

Experimental data for the T$_1$ determination
by exponential data analysis

variable delay time [sec]	δ = 44.648 signal intensity	δ = 44.575 signal intensity	δ = 16.377 signal intensity	δ = 16.304 signal intensity
0.025	-42.3	-40.9	-42.1	-41.6
0.05	-42.2	-40.6	-42.6	-42.0
0.1	-40.1	-38.5	-40.5	-39.5
0.2	-39.1	-37.2	-39.0	-37.1
0.4	-36.9	-35.2	-35.2	-33.5
0.8	-33.5	-31.8	-28.0	-26.0
1.6	-25.9	-24.6	-14.2	-13.1
3.2	-13.0	-12.4	6.58	5.83
6.4	5.57	5.33	29.2	26.3
12.8	28.6	27.0	45.5	41.5
25.6	44.8	42.3	50.4	46.5
51.2	48.6	45.8	49.4	44.9
102.4	48.9	46.2	49.6	45.9

Problem 39 : C₄H₄N₂S₂
IR: 1571, 1610, 3456 cm⁻¹
600 MHz , solvent : DMSO-d6
¹H spectrum (aromatic region) , ¹³C spectrum

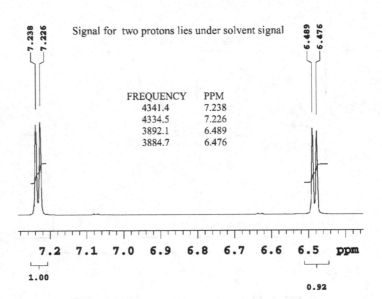

Signal for two protons lies under solvent signal

FREQUENCY	PPM
4341.4	7.238
4334.5	7.226
3892.1	6.489
3884.7	6.476

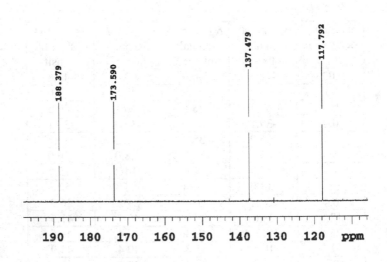

Problem 39 : $C_4H_4N_2S_2$
600 MHz , solvent : DMSO-d6
C,H and C,H long range correlation spectra

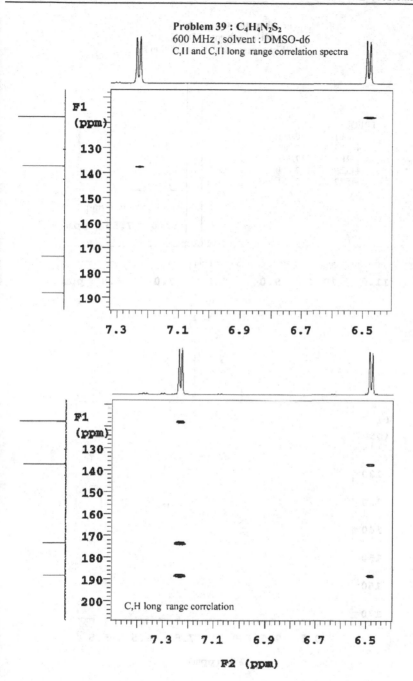

C,H long range correlation

F2 (ppm)

Problem 40 : C₅H₃Cl₂NO
600 MHz , solvent : CD₃CN
¹H spectrum and C,H correlation spectra

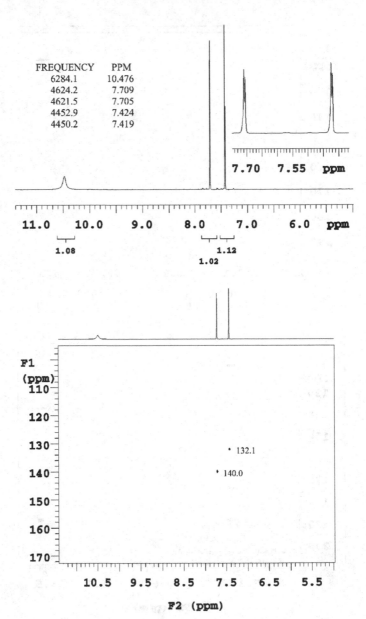

FREQUENCY	PPM
6284.1	10.476
4624.2	7.709
4621.5	7.705
4452.9	7.424
4450.2	7.419

Problem 40 : C₅H₃Cl₂NO
600 MHz , solvent : CD3CN
C,H longe rang correlation spectra
recorded using different acquisition parameters

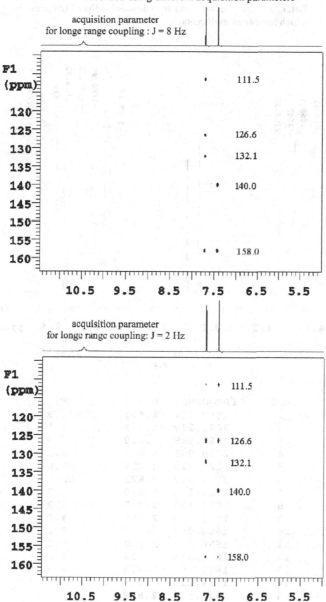

acquisition parameter
for longe range coupling : J = 8 Hz

F1
(ppm)

111.5

120
125
130
135
140
145
150
155
160

126.6

132.1

140.0

158.0

10.5   9.5   8.5   7.5   6.5   5.5

acquisition parameter
for longe range coupling: J = 2 Hz

F1
(ppm)

111.5

120
125
130
135
140
145
150
155
160

126.6

132.1

140.0

158.0

10.5   9.5   8.5   7.5   6.5   5.5

Problem 41 : $C_6H_{10}O_5$
600 MHz , solvent : $CDCl_3$
^1H spectrum and table with chemical shifts

Determine the structure and the chemical shifts of the protons
which appear as multiplets.

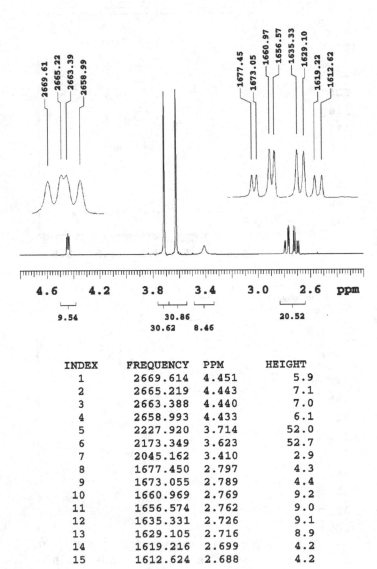

INDEX	FREQUENCY	PPM	HEIGHT
1	2669.614	4.451	5.9
2	2665.219	4.443	7.1
3	2663.388	4.440	7.0
4	2658.993	4.433	6.1
5	2227.920	3.714	52.0
6	2173.349	3.623	52.7
7	2045.162	3.410	2.9
8	1677.450	2.797	4.3
9	1673.055	2.789	4.4
10	1660.969	2.769	9.2
11	1656.574	2.762	9.0
12	1635.331	2.726	9.1
13	1629.105	2.716	8.9
14	1619.216	2.699	4.2
15	1612.624	2.688	4.2

Problem 41 : C$_6$H$_{10}$O$_5$
600 MHz , solvent : CDCl$_3$
NOESY 1D , ^{1}H and ^{13}C spectra
arrows denote signal irradiated

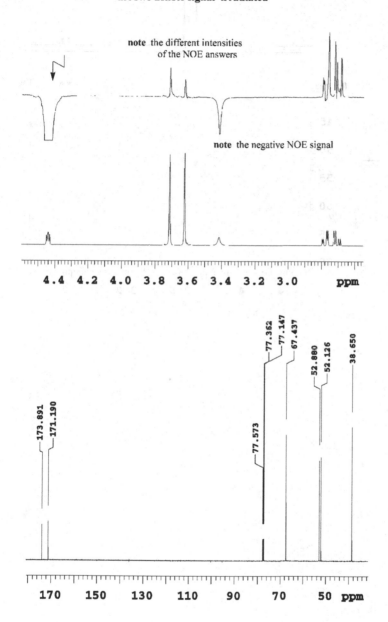

note the different intensities
of the NOE answers

note the negative NOE signal

Problem 41 : $C_6H_{10}O_5$
600 MHz , solvent : CDCl$_3$
C,H and C,H long range correlation spectra

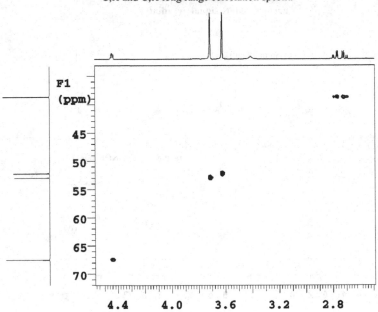

Problem 41 : C$_6$H$_{10}$O$_5$
600 MHz , solvent : CDCl$_3$
C,H and C,H long range correlation spectra

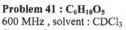

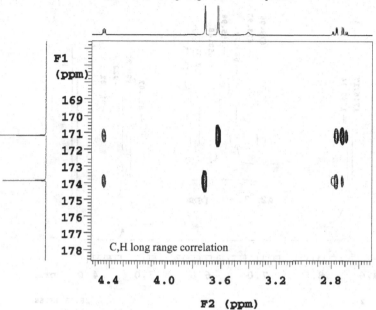

C,H long range correlation

F1 (ppm)

F2 (ppm)

Problem 42 : C₆H₁₂ClNO₄

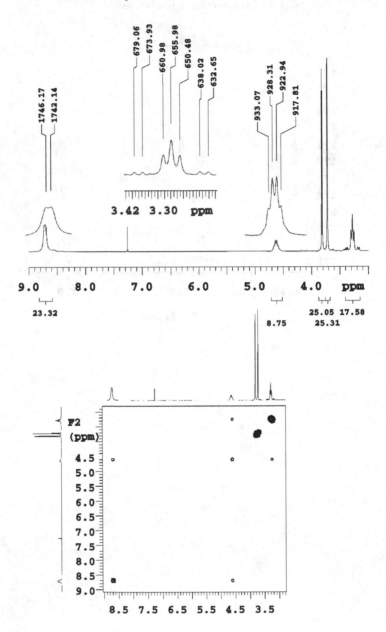

200 MHz , solvent : CDCl₃
¹H and H,H correlation spectra

Problem 42 : $C_6H_{12}ClNO_4$
200 MHz , solvent : $CDCl_3$
^1H spectrum and homodecoupling spectra
Determination of the coupling constants

Hz	PPM	Hz	PPM
930.1	4.65	682.7	3.41
925.1	4.62	677.8	3.39
920.1	4.59	664.8	3.32
		659.5	3.30
		653.4	3.27
		641.1	3.20
		635.5	3.18

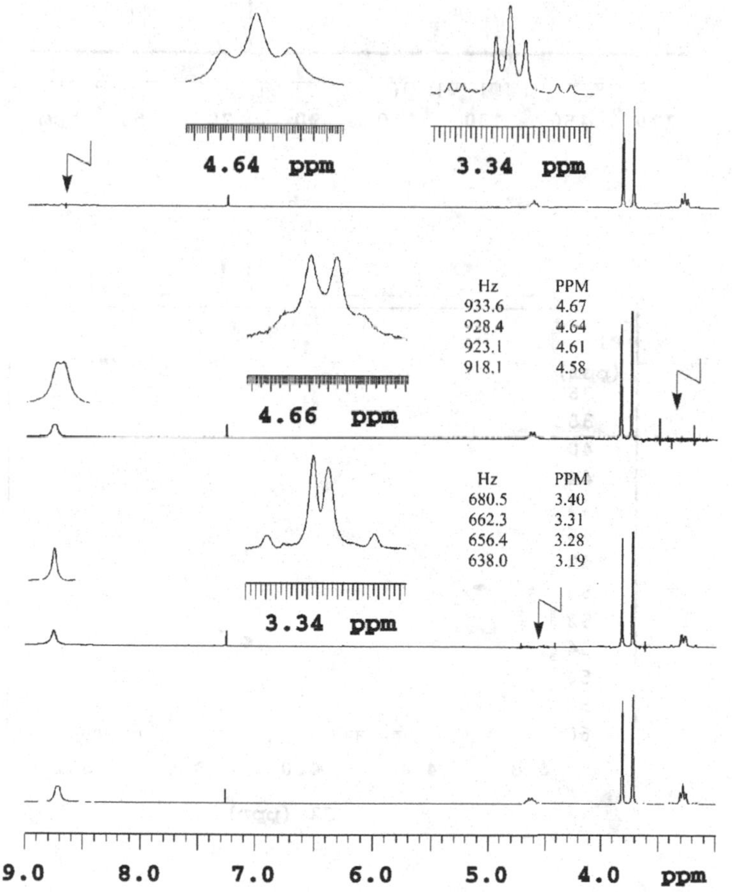

4.64 ppm   3.34 ppm

Hz	PPM
933.6	4.67
928.4	4.64
923.1	4.61
918.1	4.58

4.66 ppm

Hz	PPM
680.5	3.40
662.3	3.31
656.4	3.28
638.0	3.19

3.34 ppm

9.0   8.0   7.0   6.0   5.0   4.0   ppm

Problem 42 : C₆H₁₂ClNO₄
600 MHz , solvent : CDCl3
¹³C and C,H correlation spectra

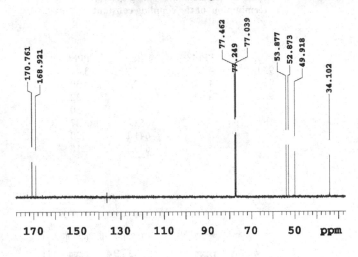

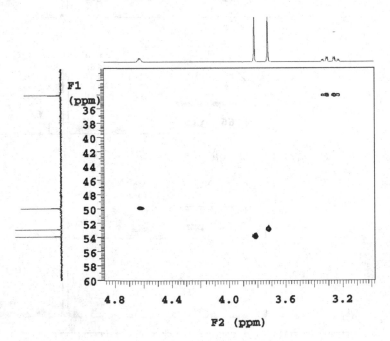

Problem 42 : C₆H₁₂ClNO₄
600 MHz , solvent : CDCl₃
parts of C,H long range correlation spectrum

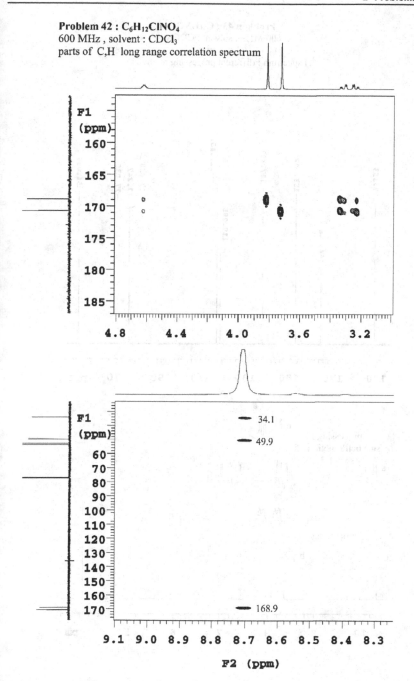

Problem 43 : C₇H₇NO₂
600 MHz , solvent : CDCl₃
¹³C spectrum
¹H spectrum : different processing methods

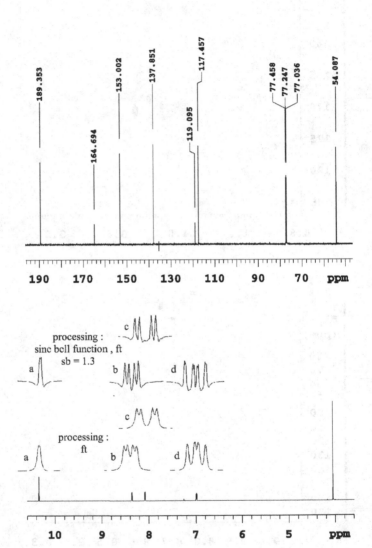

Problem 43 : C$_7$H$_7$NO$_2$
600 MHz , solvent : CDCl$_3$
CH-and CH-long range correlation spectra

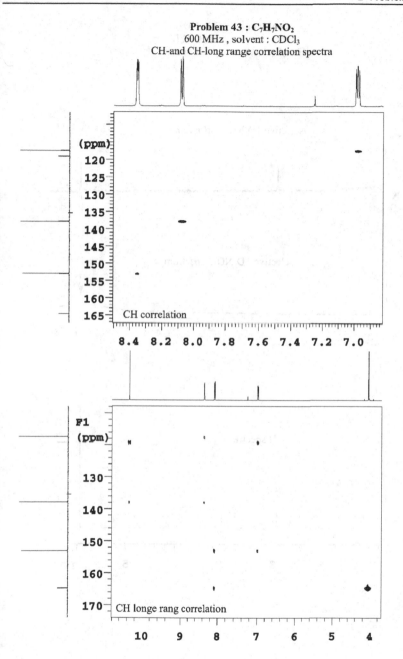

Problem 43 : C₇H₇NO₂
600 MHz , solvent : CDCl₃
¹H and selective 1D NOE spectra

selective 1D NOE spectrum

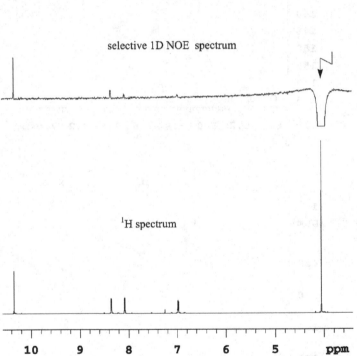

selective 1D NOE spectrum

¹H spectrum

Problem 43 : C₇H₇NO₂

$Problem\ 43 : C_7H_7NO_2$

600 MHz , solvent : CDCl₃
¹H and homodecoupling spectra
processing : with sine bell function (sb = 1)
tables with chemical shift in ppm and Hz
for the determination of coupling constants

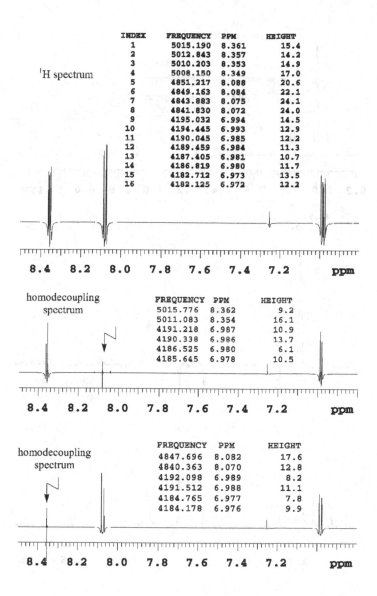

¹H spectrum

INDEX	FREQUENCY	PPM	HEIGHT
1	5015.190	8.361	15.4
2	5012.843	8.357	14.2
3	5010.203	8.353	14.9
4	5008.150	8.349	17.0
5	4851.217	8.088	20.6
6	4849.163	8.084	22.1
7	4843.883	8.075	24.1
8	4841.830	8.072	24.0
9	4195.032	6.994	14.5
10	4194.445	6.993	12.9
11	4190.045	6.985	12.2
12	4189.459	6.984	11.3
13	4187.405	6.981	10.7
14	4186.819	6.980	11.7
15	4182.712	6.973	13.5
16	4182.125	6.972	12.2

```
   8.4   8.2   8.0   7.8   7.6   7.4   7.2        ppm
```

homodecoupling spectrum

FREQUENCY	PPM	HEIGHT
5015.776	8.362	9.2
5011.083	8.354	16.1
4191.218	6.987	10.9
4190.338	6.986	13.7
4186.525	6.980	6.1
4185.645	6.978	10.5

```
   8.4   8.2   8.0   7.8   7.6   7.4   7.2        ppm
```

homodecoupling spectrum

FREQUENCY	PPM	HEIGHT
4847.696	8.082	17.6
4840.363	8.070	12.8
4192.098	6.989	8.2
4191.512	6.988	11.1
4184.765	6.977	7.8
4184.178	6.976	9.9

```
   8.4   8.2   8.0   7.8   7.6   7.4   7.2        ppm
```

Problem 44 : $C_8H_3BrF_6$
^{19}F NMR: $\delta = -63.7$ ppm
200 MHz , solvent : $CDCl_3$
^1H , APT and ^{13}C spectra

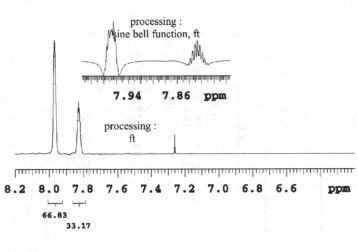

processing :
sine bell function, ft

7.94   7.86   ppm

processing :
ft

8.2   8.0   7.8   7.6   7.4   7.2   7.0   6.8   6.6   ppm

66.83
33.17

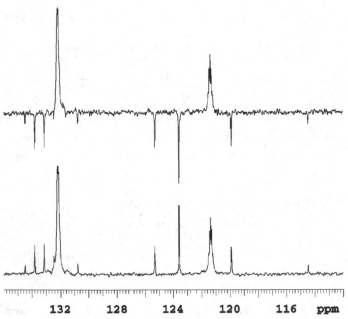

132   128   124   120   116   ppm

Problem 44 : C₈H₃BrF₆
200 MHz , solvent : CDCl₃

Table: ¹³C chemical shifts (¹H decoupled)
(in ppm and Hz : for the determination of coupling constants)

INDEX	FREQUENCY	PPM	HEIGHT
1	6767.282	134.479	3.1
2	6733.157	133.800	9.4
3	6699.032	133.122	9.8
4	6664.907	132.444	5.8
5	6652.030	132.188	35.0
6	6648.810	132.124	34.8
7	6579.272	130.742	3.4
8	6306.271	125.317	9.0
9	6219.349	123.590	22.2
10	6110.535	121.428	14.0
11	6106.672	121.351	18.4
12	6102.808	121.274	14.4
13	6033.270	119.892	8.8

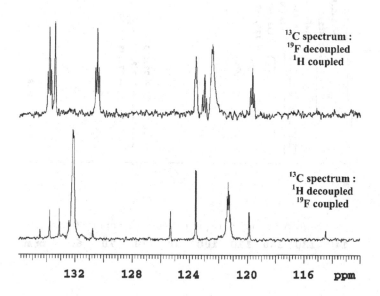

¹³C spectrum :
¹⁹F decoupled
¹H coupled

¹³C spectrum :
¹H decoupled
¹⁹F coupled

132   128   124   120   116   ppm

Problem 45 : C₉H₇Br
600 MHz , solvent : CDCl₃
¹H and ¹³C APT spectra

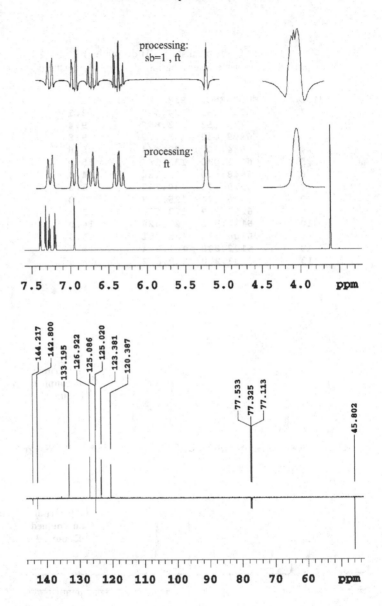

Problem 45: C₉H₇Br
600 MHz , solvent : CDCl₃
¹H and TOCSY-1D and NOESY-1D spectra

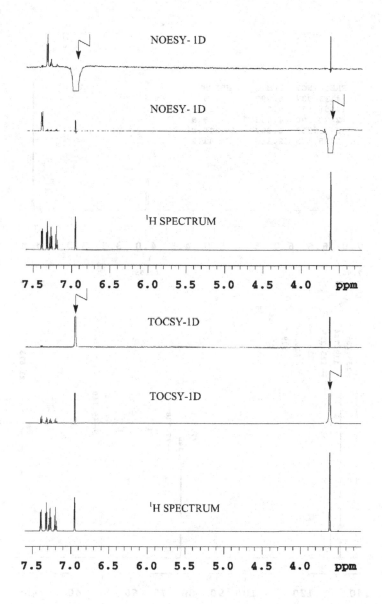

Problem 46 : C9H11Br2P
^{31}P NMR: δ = 173.6 ppm
600 MHz , solvent : CDCl₃
^{1}H and ^{13}C spectra

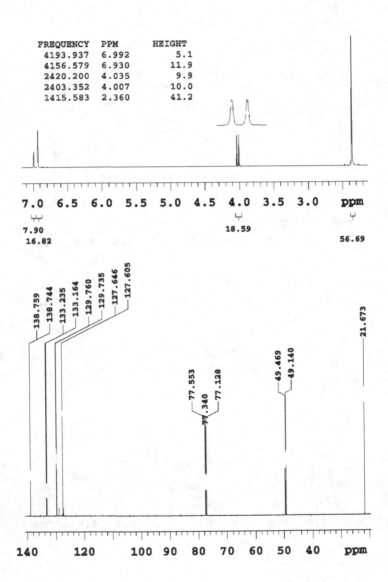

FREQUENCY	PPM	HEIGHT
4193.937	6.992	5.1
4156.579	6.930	11.9
2420.200	4.035	9.9
2403.352	4.007	10.0
1415.583	2.360	41.2

7.0   6.5   6.0   5.5   5.0   4.5   4.0   3.5   3.0   ppm

7.90
16.82   18.59   56.69

138.759
138.744
133.235
133.164
129.760
129.735
127.646
127.605

77.553
77.340
77.128

49.469
49.140

21.673

140   120   100 90 80 70 60 50 40   ppm

Problem 46 : C$_9$H$_{11}$Br$_2$P
600 MHz , solvent : CDCl$_3$
NOESY spectrum

Table : ^{13}C chemical shifts
(in ppm and Hz : for the determination of coupling constants)

FREQUENCY	PPM	HEIGHT
20928.672	138.759	11.3
20926.383	138.744	9.6
20095.542	133.235	5.7
20084.861	133.164	5.4
19571.403	129.760	16.6
19567.588	129.735	14.7
19252.494	127.646	28.4
19246.391	127.605	31.5
11697.105	77.553	8.1
11665.061	77.340	8.1
11633.018	77.128	7.9
7461.265	49.469	14.7
7411.674	49.140	15.6
3268.913	21.673	35.3

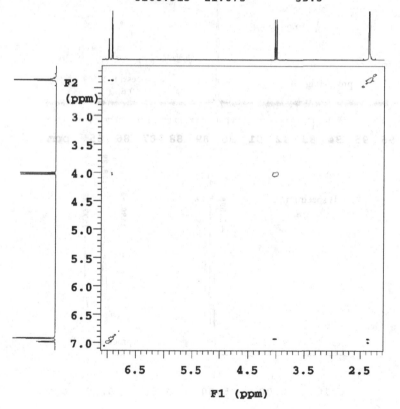

Problem 47 : C₁₀H₁₀F₂Ti
200 MHz , solvent : C₆D₆
¹⁹F and ¹H spectra

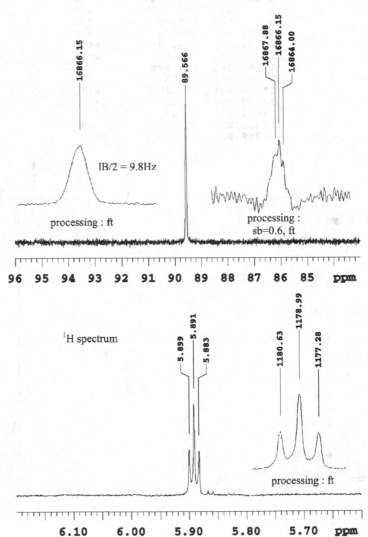

¹⁹F spectrum
(¹H coupled, expansions using different processing methods)

Problem 47: $C_{10}H_{10}F_2Ti$
200 MHz , solvent : C_6D_6
^{13}C and C,H correlation spectra

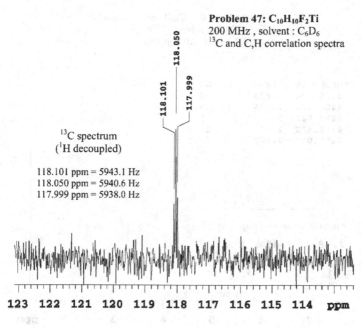

^{13}C spectrum
(^1H decoupled)

118.101 ppm = 5943.1 Hz
118.050 ppm = 5940.6 Hz
117.999 ppm = 5938.0 Hz

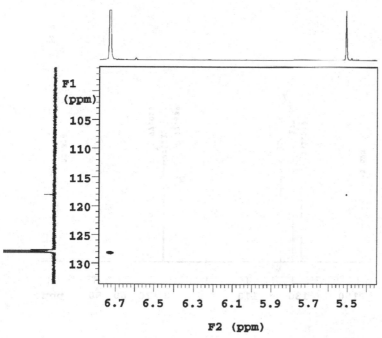

Problem 48 : $C_{10}H_{12}Br_2$
600 MHz , solvent : CDCl$_3$
^{1}H and ^{13}C APT spectra

FREQUENCY	PPM	HEIGHT
4485.836	7.478	4.4
4484.371	7.476	9.0
4482.540	7.473	5.8
4456.902	7.430	21.2
4455.071	7.427	20.7

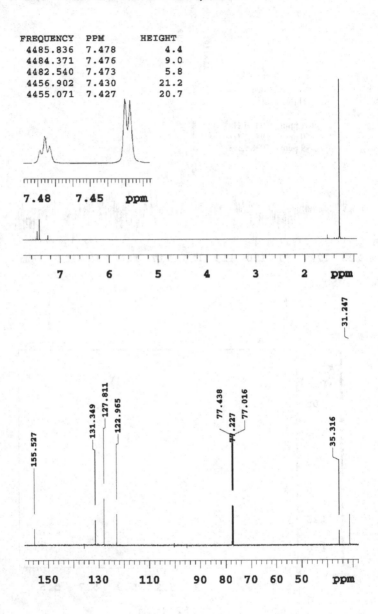

Problem 48 : C₁₀H₁₂Br₂

$Problem\ 48 : C_{10}H_{12}Br_2$

600 MHz , solvent : CDCl₃

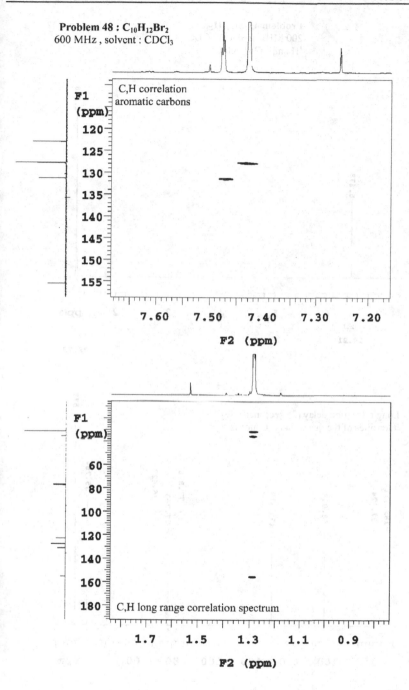

F1
(ppm)

C,H correlation
aromatic carbons

120
125
130
135
140
145
150
155

7.60   7.50   7.40   7.30   7.20

F2 (ppm)

F1
(ppm)

60
80
100
120
140
160
180

C,H long range correlation spectrum

1.7   1.5   1.3   1.1   0.9

F2 (ppm)

Problem 49 : C₁₄H₂₀O₂
200 MHz , solvent : CDCl₃
¹H and ¹³C spectra

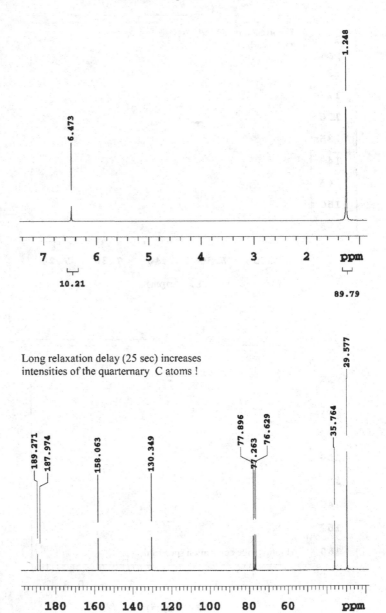

Long relaxation delay (25 sec) increases
intensities of the quarternary C atoms !

Problem 49 : C$_{14}$H$_{20}$O$_2$
600 MHz , solvent : CDCl$_3$

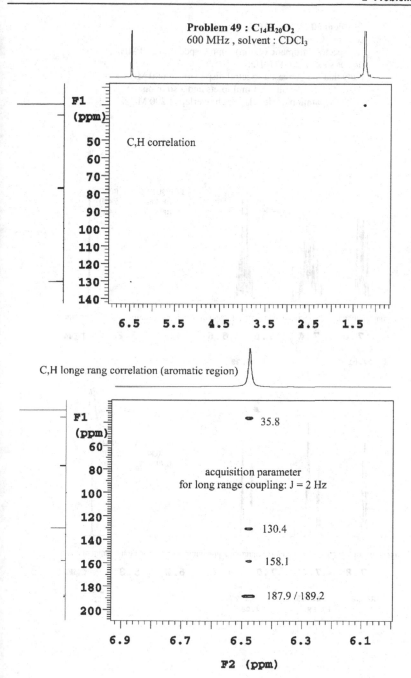

C,H correlation

C,H longe rang correlation (aromatic region)

F1
(ppm)
60

acquisition parameter
for long range coupling: J = 2 Hz

35.8

130.4

158.1

187.9 / 189.2

F2 (ppm)

Problem 50 : C₁₉H₁₅OPS₂

solvent : CDCl₃

¹H spectra (aromatic region) : upper spectrum 200 MHz ,
lower spectrum 600 MHz.

Note that the increased spectral dispersion at 600 MHz
leads to less complicated multiplets and also in one case
to the separation of signals which overlap at 200 MHz.

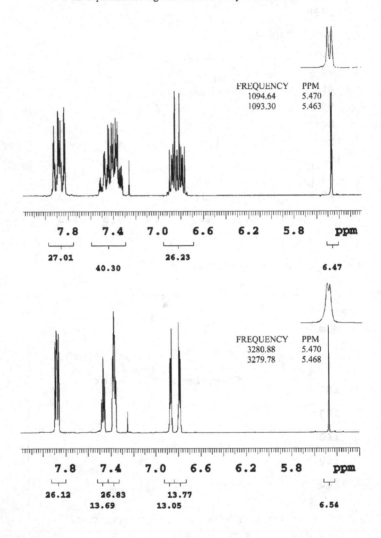

FREQUENCY   PPM
1094.64   5.470
1093.30   5.463

7.8   7.4   7.0   6.6   6.2   5.8   ppm

27.01   26.23
       40.30   6.47

FREQUENCY   PPM
3280.88   5.470
3279.78   5.468

7.8   7.4   7.0   6.6   6.2   5.8   ppm

26.12   26.83   13.77
         13.69   13.05   6.54

Problem 50 : C$_{19}$H$_{15}$OPS$_2$
solvent : CDCl$_3$
^{1}H spectra (aromatic region) : 200 MHz (top) , 600 MHz (below)
note the different fine structures and dispersion

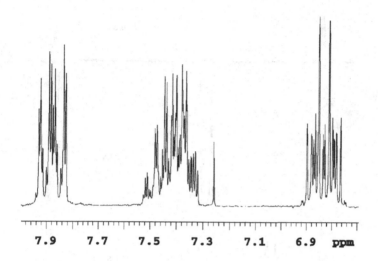

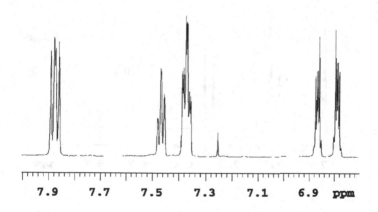

Problem 50 : C_{19}H_{15}OPS_2
600 MHz , solvent : CDCl_3
selective NOESY 1D, TOCSY 1D and ¹H spectra

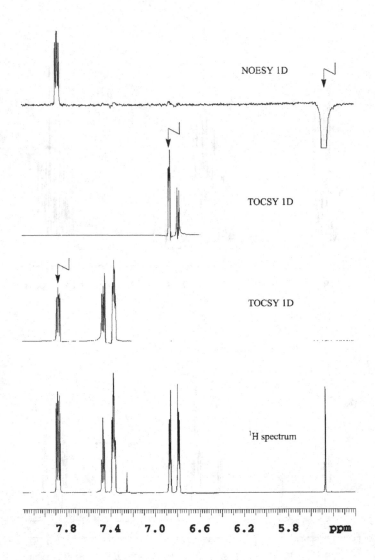

NOESY 1D

TOCSY 1D

TOCSY 1D

¹H spectrum

7.8   7.4   7.0   6.6   6.2   5.8   ppm

Problem 50 : C₁₉H₁₅OPS₂
600 MHz , solvent : CDCl₃
¹³C spectrum and table with chemical shifts
(for the coupling constants)

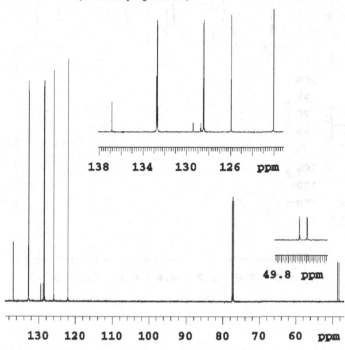

INDEX	FREQUENCY	PPM	HEIGHT
1	20640.091	136.846	19.3
2	20022.830	132.753	33.8
3	20019.951	132.734	35.7
4	20014.193	132.696	71.3
5	20005.556	132.639	68.2
6	19519.002	129.413	5.7
7	19417.661	128.741	5.9
8	19383.112	128.512	69.4
9	19371.596	128.435	71.5
10	18995.597	125.942	74.8
11	18405.398	122.029	78.3
12	11689.804	77.504	32.6
13	11657.559	77.291	34.1
14	11625.890	77.081	32.6
15	7329.245	48.594	12.7
16	7263.028	48.154	12.5

Problem 50 : C₁₉H₁₅OPS₂
600 MHz , solvent : CDCl₃
C,H and C,H long range correlation spectra

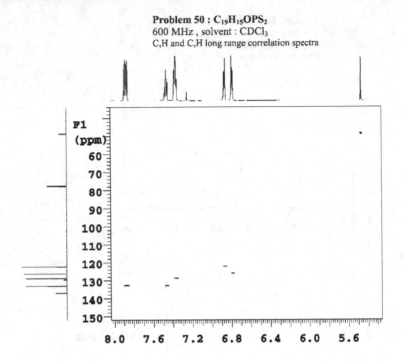

Problem 50 : C$_{19}$H$_{15}$OPS$_2$
600 MHz , solvent : CDCl$_3$
C,H and C,H long range correlation spectra

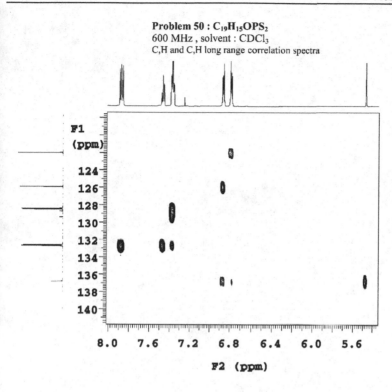

Printed in the United States
By Bookmasters